D1567597

ON BECOMING A PROFESSIONAL GEOGRAPHER

Edited by

MARTIN S. KENZER
Louisiana State University

MERRILL PUBLISHING COMPANY
A Bell & Howell Information Company
Columbus Toronto London Melbourne

Published by Merrill Publishing Company
A Bell & Howell Information Company
Columbus, Ohio 43216

This book was set in Garamond.

Administrative Editor: Wendy W. Jones
Production Coordinator: Linda Kauffman Peterson
Art Coordinator: Vincent A. Smith
Cover Designer: Brian Deep
Text Designer: Anne Daly

Library of Congress Catalog Card Number: 89-60329 (Casebound)
89-60310 (Paperback)
International Standard Book Number: 0-675-20679-0 (Casebound)
0-675-20680-4 (Paperback)
Printed in the United States of America
1 2 3 4 5 6 7 8 9—92 91 90 89

Contributors

Ronald F. Abler
Department of Geography
Pennsylvania State University

John A. Alwin
Department of Geography
Dartmouth College

L. S. Bourne
Department of Geography
University of Toronto
Canada

David R. Butler
Department of Geography
University of Georgia

John E. Chappell, Jr.
San Luis Obispo, California

H. J. de Blij
Department of Geography
University of Miami

Larry R. Ford
Department of Geography
San Diego State University

Leonard T. Guelke
Department of Geography
University of Waterloo
Canada

David Hornbeck
Department of Geography
California State University-Northridge

John A. Jakle
Department of Geography
University of Illinois

Martin S. Kenzer
Department of Geography and
Anthropology
Louisiana State University

Bonnie S. Loyd
Landscape
Berkeley, California

Robert A. Muller
Department of Geography and
Anthropology
Louisiana State University

Risa Palm
Graduate School
University of Colorado

Douglas B. Richardson
GeoResearch, Inc.
Billings, Montana

Robert A. Rundstrom
Department of Public Affairs/Geography
George Mason University

Barney Warf
Port Authority of NY–NJ
New York, New York

William Wyckoff
Department of Earth Sciences
Montana State University

Preface

The completion of *On Becoming a Professional Geographer,* and its production by Merrill Publishing Company, came to pass because of the kind assistance and sincere encouragement of several individuals. I thank Sam Hilliard and Carville Earle, in the Department of Geography and Anthropology at Louisiana State University, for their support. Special appreciation goes to some key people at Merrill: Wendy W. Jones for her perceptiveness; Linda Kauffman Peterson and Diane Jordan for their fine-tuning skills, their patience, and their sympathetic personalities; and the other members of the Merrill team, with whom it has been a pleasure to work. Likewise, at the University of Southern California (where I was a visitor during the spring 1989 semester), I also thank Curtis C. Roseman and the rest of the Geography Department for providing me with a cordial environment in which to attend to the last-minute details of the book.

Finally, the individual authors of the chapters deserve special praise. They abided by my demanding schedules and still managed to produce something for which they can be proud. All volume editors should be so fortunate.

Martin S. Kenzer
Los Angeles, California

Contents

INTRODUCTION

On the Origins and Purposes of This Book

Martin S. Kenzer

One of my initial thoughts about graduate school, curiously enough, was no thought at all. About the time I was finishing my B.A., an instructor asked me where I intended to pursue my graduate degree. I remember that the question elicited a total blank on my part; I really had no conception of what graduate school was all about, much less which one I might attend. I didn't know what to expect of a graduate education, how many years it would take, or, frankly, that I had such an option. I had assumed that I would simply get a job after the four-year undergraduate degree. More important, however, I recall thinking that graduate school was only for those who wanted to teach or to make a lot of money as a medical doctor, lawyer, engineer, or business executive. Graduate school was a foreign concept to me, and I shied away from it for a number of years after my baccalaureate.

When I finally decided to pursue a graduate degree, I vividly remember being scared to death at the prospect of my first *seminar* meeting—the word itself had me envisioning and anticipating all sorts of horrible fates. I somehow survived all the new experiences, but it was truly "on-the-job training." I remember pretending that each new experience was normal, nothing to get worked up about. With my M.A. behind me, I went on for my Ph.D., again not knowing what to expect, trying not to ask too many questions, and hoping not to look too scared or too ignorant. As my doctoral program proceeded, although little was said aloud, I gradually realized that I wasn't alone: My peers were all in the same boat, all playing a similar game, learning what was expected of them by trial and error. Eventually, those who persevered, did what they were told, and asked the fewest questions about procedure, finished; those who questioned everything generally gave in to their confusion and frustration, and many of them dropped out. In some cases, decidedly intelligent, self-motivated students were forced out of the system mainly because they were unable to adapt to a "game" that seemingly had few or no rules. At the time, it seemed as though the "rules" were intentionally vague and often contradictory.

Ironically, while graduate school teaches you how to do a good many things, it rarely, and only implicitly, alerts you to the fact that there is no one right way to do them—just as there is no one "right" way of getting through graduate school.

You do your best, and you either pass or fail. Likewise, getting through the system successfully does not guarantee that you will then be able to "make it" after your graduate education. Just as paroled criminals quickly discover that rehabilitation programs mean nothing if they are not taught how to apply their new knowledge to the outside world, graduate training imparts knowledge but says very little about what to do with that knowledge once you have finished your degree. In a sense, you "win" the game without ever having learned the rules. Often success seems to come from simple perseverance.

Long before I finished my formal education, it occurred to me that I was not being properly trained for life beyond graduate school. I had learned a great many techniques: how to use a library; how to organize vast quantities of data effectively; how to recognize the differences between primary and secondary research; how to apply statistics; how to analyze and interpret information geographically; how to construct a term paper, thesis, or dissertation; how to bolster an argument with sources; which academic journals and which academicians were important to cite; how to think critically; and, of course, how to find fault with the work of others. What I had not learned, however, was a practical way to put these skills to use, to apply my geographic education to the workaday academic world; and this was even more true with respect to the nonacademic world. I watched instructors for clues, but I had no "inside" look at what they did, what was expected of them, or how they tackled everyday concerns.

I doubt that this was because of the particular universities I attended. For the most part, in retrospect, they were excellent, and, in some ways, I was extremely fortunate in this regard. I have seen others pass through graduate school with far inferior training than I received. A significant number seem to finish their degrees despite their poor training. Some then manage to survive after years of struggling and scores of movie-like experiences; others complete graduate school, but, for a variety of reasons, receive virtually no practical, real world education; the latter typically struggle even more than the former.

For graduate students at any university, entering the post-graduate school world can be either intriguing or scary. It is of course a welcome relief to be *finally* out of school: never to have to take another exam—or so you believe; to be rid at long last of the daily and weekly schedules of academia—or so it seems at the time; not to have to worry about reading assignments, term papers, and grades. Yet, one soon discovers that things don't change quite so dramatically. Beyond graduate school schedules persist, and often they get worse. You continue to take different types of exams, and, although the contexts change, you most assuredly continue to compete with colleagues. And, unless you wind up doing purely manual labor, you will still have to produce written assignments.

At a basic level, getting a job—particularly the first one—in any field can be frightening. What does it take to get that initial position, and once you have it, what then? What can you anticipate? How will it differ from your student days? After what seems like, and in some cases is, a lifetime of going to school, how do you prepare for a productive existence beyond the classroom? How is it possible to prepare for a career outside of academia? If you decide to become a teacher, will you be prepared to assume a whole new realm of responsibilities? What, exactly, will those new responsibilities be?

If you're completing a baccalaureate degree, what can you expect of graduate school? How will it differ from the previous four years? If you're about to leave graduate school, now what? How do you prepare for the post-graduate school world? In short, how do you survive *effectively* during and beyond graduate school?

It was with these questions in mind that I organized this volume. I understood from the outset, however, that not every important issue or crucial question could be addressed in these pages.[1] Most of the chapters are based on my perception of (a) the questions that new and potential geography graduate students might ask, (b) the most important issues graduate students in geography face as they approach the end of their formal training, and (c) the immediate worries of new faculty members. If the emphasis is on the academic world beyond graduate school, it reflects the issues and concerns I know best. A different volume editor naturally might have selected different topics to discuss. Further, some readers may find certain chapters helpful but other chapters of no immediate value. Some may agree with many or all of the essays, while others will have different views. This, too, was expected.

The book is divided into five parts, each addressing different aspects of graduate and post-graduate school life. The four chapters of Part I, "On Graduate School," answer some fundamental questions about the graduate-school experience.

In Chapter 1, David Hornbeck presents his views on the nature of graduate education—how it differs from the undergraduate experience and what a prospective student should anticipate—as well as the steps involved in selecting the "right" school. He argues that a graduate education is *not* for everyone, and that if you opt to go you must be prepared to expend the required effort or you will be wasting everyone's time.

In Chapter 2, John Chappell, Jr. asks both new and practicing geographers to reconsider the discipline in light of its so-called uniqueness. He suggests we regard our subject matter as "the surface of the earth," and that we divorce ourselves from strict, limiting definitions. A second issue of concern for Chappell is our history of indiscriminate borrowing from other disciplines in the name of "synthesis." Too often, he notes, we borrow secondhand material and thereby lose precious insight into the nature of the borrowed ideas. Chappell's essay carries an important message for all geographers, irrespective of subfield or specialty.

Chapters 3 and 4 address practical aspects of a graduate education. Larry Ford describes right and wrong ways to deliver an oral presentation. If you intend to become a professional geographer, you will, eventually, be asked to present your research orally: at a local, regional, or national conference; as an instructor in front of a class; as a research associate at a meeting of interested clients; as a government employee asked to address the public on some topic of general concern; or in a dozen other possible situations. Both novice and experienced geographers alike will profit from the cogent comments in this paper. Chapter 4, by William Wyckoff, likewise will be helpful for geographers spanning the entire academic employment range, from new graduate student to full professor. His paper on job searches is pertinent to all of us, for few geographers remain at the same university for very

[1] In case this book is ever revised or updated, I would appreciate hearing from readers who feel that critical topics are missing.

long. Graduate students who have never gone through the ordeal of an academic job interview will find his words of wisdom of immediate value, while those who have "been there" can certainly pick up a few pointers, too. As Wyckoff correctly points out, the bottom line is to anticipate your audience and take charge of the situation whenever possible.

Part II, "Beyond Graduate School," will be of interest to graduate students pondering the work force and life beyond their degree program, as well as to those considering the possibility of changing careers. I asked the three authors to describe "the transition" from graduate school to the so-called real world—no small task, indeed, yet many of their comments substantially transcend those instructions. In Chapter 5, Robert Rundstrom reveals some of the tensions, anxieties, and frustrations of newly appointed instructors. Caught in a downward spiral of never-ending work and commitments, new instructors struggle to keep their heads above water—sometimes barely above. Eventually, they find themselves rising up and out of the whirlwind. Somewhat tongue-in-cheek, the chapter is nonetheless full of perceptive insights into the neophyte professor's lifestyle. Douglas Richardson, in Chapter 6, carries the discussion from academia into the private sector. A geographer who owns a consulting firm and thus engages in very practical, real-world endeavors, Richardson says a great deal about the state of the discipline and the supposed division between applied and theoretical research. As he points out, while there are no doubt distinctions between the university and the private sector, the same criteria of quality and common sense apply equally to both. Richardson offers a number of timely points to contemplate that should prove useful for any geographer thinking about working for a private company. Finally, Barney Warf in Chapter 7 writes about opportunities for geographers in the government sector. He has a particularly good vantage point on this topic, since he has taught at the university level but opted to leave academia for government employment. He concludes that this sector of the work force is an ideal but often overlooked setting for geographers, and that the potential for geographical research in government is unlimited. As Warf so aptly notes, those who are socially minded will find that working for the government places them squarely in the midst of the planners and the decision-makers who ultimately will determine future socio-economic conditions.

Professional geographers, regardless of position and type of work, almost certainly will be required to write—and write, and write. There are very few geographers working in a professional capacity who aren't asked to produce written reports of some kind, generally on a regular basis. Accordingly, Part III, "On Writing," is devoted to the varied aspects of the writing assignment: writing scholarly articles, in both physical and human geography, writing for the public, writing books for one's peers, and—what some would claim to be the most difficult and time-consuming task of all—writing geography textbooks.

Chapters 8 and 9, by David Butler and Larry Bourne, respectively, address the question every academic geographer asks at least once: "How do I get my articles published?" Although there are no hard and fast rules that guarantee success, Butler and Bourne suggest surprisingly similar guidelines. Both examine the problems of coauthorship—an issue more germane to physical geographers—and note common problems with regard to editorial decisions. If there is a common denominator to their wisdom, however, it is perseverance—no one gains success overnight, and the production of scholarly journal articles is no exception to that rule. Also, both remind us to keep our intended audience in mind at all times and not delude ourselves into thinking a journal will change just for us.

Chapters 10–12, by John Alwin, John Jakle, and Harm de Blij, respectively, also call attention to the importance of gauging one's anticipated audience. Alwin notes how the average nonacademic reader responds to, or more typically is turned off by, the jargon that all too often fills the pages of our journals. When writing for the public, he says, keep it simple but evocative; treat your readers with the respect they deserve or you will rapidly lose their interest.

In Chapter 11, Jakle offers some helpful tips on writing scholarly books and dealing with presses. A prolific book author, Jakle offers insights into the world of scholarly publishing that should be heeded by anyone who plans to write a book. Of particular relevancy are his comments on book reviewing, an act of "service" that most professional geographers will be asked to perform at some point, and that far too many take for granted.

Textbooks pose peculiar problems, and certainly not every geographer will attempt to write one, nor should most even try. The academic rewards are small, but how would the discipline function without them? We all use texts in our classrooms, and we are all quick to point out their weaknesses. Yet very few of us will ever commit ourselves to the time-consuming and demanding tasks of writing and revision that such projects entail. But, for those who feel inclined to make the commitment, de Blij's remarks on textbook production will be required reading. He may be the most accomplished author of geography textbooks, and he speaks from years of experience. In Chapter 12, he discusses not only the genesis of his best-known texts but also several of the more demanding aspects of textbook production.

Part IV on "Editorial and Grant-Getting 'Secrets'" aims to demystify some of the most potentially frustrating and least-understood endeavors for professional geographers: dealing with journal editors, organizing multi-authored collections such as this, and obtaining external funding. Bonnie Loyd, in Chapter 13, details the editorial process from her perspective as a geographer-editor, offering an "inside" view of the editorial and manuscript review processes. She emphasizes that you must know both your audience and your prospective journal. To avoid problems, she advises, follow house style from the start. Also, as managing editor of *Landscape,* she is able to offer invaluable tips on the art of photography for publication. In Chapter 14, on compiling and editing multi-authored volumes, I deal with issues of procedure and responsibility. Such collections differ dramatically from other types of books, and those contemplating such an enterprise should be aware of the effort involved and the unique nature of the undertaking. Organizational skills are a necessity, as is the ability to generate and sustain self-motivation throughout a project that may take many months to complete, only to discover that tenure and promotion committees look askance at the final product.

Ronald Abler's contribution, Chapter 15, will be read by every geographer—no doubt several times. Using his background as director of the Geography and Regional Science Program at the National Science Foundation as his guide, he provides thoughtful hints on how to secure outside grant money. The key is to be organized and logical in your thinking and as concise as possible in your writing, but there's much more to it than that. Otherwise, as his opening quotation suggests, we would all get what we ask for. Abler offers an internal glimpse into the proposal review process. He concludes that grant-getting is a "learnable skill."

Part V, "On Academic Survival," highlights several issues faced solely by those working in the world of post-secondary education. The mechanics of "survival" at colleges and universities—at the level of the individual and also at the departmental level—are unique. Individuals must survive the tenure and promotion process, and departments, from time to time, must survive internal and external review committee decisions. Neither comes easily. Although both situations have distinctive characteristics, both can be frustrating and potentially debilitating. In Chapter 16, Leonard Guelke tackles the sensitive subject of tenure and promotion and offers intelligent guidelines to follow, both to obtain tenure and promotion and to appeal adverse decisions when necessary. As he makes abundantly clear, there are very definite roads to take in the quest for tenure and promotion, and there are equally well-known routes *not* to take. In some instances, timing can be the single most critical factor—but always be alert and always be prepared.

Tenure and promotion are generally based on a triad of accomplishments: publications, teaching, and a much more nebulous factor termed "service." While what is expected in the two former activities is generally well understood, types and degrees of service can vary widely. To help clarify matters, Robert Muller, in Chapter 17, offers a personal reflection on the relative importance of service from several different perspectives. Written with the beginning assistant professor in mind, his essay outlines those service-related endeavors that might be considered important, as well as those that might be viewed as less important. In the final analysis, teaching and publications generally supersede service, but in Muller's view that does *not* mean that service can or should be ignored.

Finally, Chapter 18 is a prescription for departmental survival. Risa Palm, writing as a geographer-administrator, describes the crucial program review process. At a time when American geography departments are being examined and scrutinized more intently than ever before, her views and comments serve as welcome recommendations to help offset future departmental crises. Palm uses the University of Colorado as her model, but her insights apply throughout the country and in the Canadian context, too. While some departments fear such reviews, Palm calls attention to their positive aspects, whereby departmental "goals and interests" can be strengthened.

The chapters in this book, by their nature, are all personal, subjective essays. As the volume editor, I constantly had to remind and reassure the authors of this fact, because graduate school trains us to be "objective" and opinionless. But this is necessarily a collection of opinionated essays. The sorts of issues and questions being considered here do not readily, if at all, lend themselves to objective, cut-and-dried answers. As a consequence, no two geographers would approach a given chapter in the same way, and many readers will probably disagree with many of the "conclusions" found in each paper. The purpose of this collection, however, is not an attempt to impose prescriptive views on these related issues. Neither I nor the contributors ever intended to convince anyone that this is the one, correct, proper way to do things; in short, this book is *not* a how-to manual. Rather, it is a first attempt by professional geographers to discuss openly some very important topics—many considered here in print for the first time—that will aid future geographers to compete better in graduate school and in the marketplace following their graduate education. It is a book *by* practicing geographers written *for* future practicing geographers. The chapter topics are not unique to geography, but they are discussed with a geographer's background and interests in mind.

When I entered graduate school, and especially later, when I started to teach, it became clear to me that I was indeed on my own. This collection, therefore, is a first step toward helping current and future American and Canadian geographers confront and address the learning process of professionalization. I have targeted the United States and Canada because their educational systems are quite similar and the post-graduate school opportunities and socio-economic conditions, particularly in higher education, are much the same. While I do not believe that conditions in the two nations are synonymous, as a product of both educational systems I do believe there are increasingly more similarities than differences. I make less claim for the book's relevancy beyond these two countries, where conditions are decidedly different.

I trust all geographers will find something to ponder in the chapters that follow. If this book has any lasting value, it will become evident in the next generation of professional geographers.

ON GRADUATE SCHOOL

So. . .You Wanna Go to Graduate School?

David Hornbeck

Geography is an exciting field and a broad discipline that encompasses many different topics. The breadth of geography is probably the reason you selected it as a major, even though your parents and all of your friends thought you were foolish because, as everyone knows, "the only thing you can do with a B.A. in geography is teach." Of course, we know this is not entirely true because many geographers hold responsible positions in government and business. Yet, there remains the nagging question of what to do after graduation, particularly when none of the available jobs interest you, and you feel that you would like to know more about geography. Many graduates in this situation think of continuing their education by going to graduate school.

If you are one of those thinking of graduate education, what should you do to go about getting into graduate school? Before I describe a procedure you might follow, allow me to suggest some rethinking on your part.

Graduate education is the pursuit of knowledge for its own sake, and research is thus the very foundation of graduate school. Students sometimes enter graduate school for the wrong reasons. A person who is very bright is often encouraged to go. Some students may be interested in graduate school because they want to teach, while others want a graduate degree for economic betterment.

Graduate school may not be right for you, however, even if you are very bright. Graduate school can be a straitjacket for an imaginatively endowed mind; it can stifle creativity. Mozart, Dostoevsky, Joyce, and Stravinsky, for example, all excelled in areas far removed from the kinds of traditions and environment that stem from graduate school. For many people, a graduate education is like trying to put a square peg into a round hole: it does not fit. Many of today's most respected public figures, business executives, and artists have never set foot in graduate school.

For many students, graduate school may appear to be the means by which they can enter the classroom as teachers. Beware, however, for many graduate schools do not train students to teach; rather, they concentrate on research. Research is the reason for their existence, and graduate students are therefore fully imbued with the research tradition. This creates an odd situation, because most academics who teach in universities were trained in graduate school to become researchers, not teachers. A good researcher is not necessarily a good teacher. Few, if any, graduate schools offer courses on how to teach. In essence, most university professors are doing a job for which they were never trained.

If you have a keen interest in teaching, keep in mind that teaching is an art form—it can be done brilliantly or appallingly, most frequently the latter. Teaching a class, meeting after meeting, year after year, is as much an art form as writing a series of successful short stories, or consistently painting great art. Imagine the agony of the genuinely knowledgeable teacher who does not teach well. Teaching is not for everyone, no matter how knowledgeable they may be about the subject. I am sure that most professionals outside of academia, who have gained an immense amount of knowledge from years of experience, would not be able to offer a good course. If your intention is to pursue a graduate degree to become a teacher, you should take a few moments to ask yourself (a) whether you have the talent to teach, and (b) whether graduate school is the most appropriate path to your goal.

For those of you who hope that going on to graduate school will improve your prospects for a job in government or business, it may or may not be to your economic advantage. Graduate students typically take two to three years to obtain an M.A. and another two to five years for the Ph.D. If you seek an advanced degree in hopes of a better paying job, time is not on your side. By the time you finish your graduate degree(s), those who chose to enter the job market without an advanced degree will have earned considerably more money than a graduate assistantship pays, and will have acquired two to five years of valuable on-the-job experience. Many companies consider job experience more valuable than a graduate education. Unless you have a specific job in mind that can be obtained only with an advanced degree, I would suggest that you seriously reconsider the economics of a graduate education.

If, after reading the previous paragraphs, you now realize that graduate school is not for you, then you need not read further. If, however, after reading the previous paragraphs, you are not discouraged by either the economic prospects of a graduate degree or the difficulties of teaching, you should ask yourself the question, "Why do I want to go to graduate school?" If your answer is because you thoroughly enjoy geography and want to know more about the subject, then graduate school may be for you.

One last point: If you decide to go to graduate school, remember that you are entering not a democratic institution but rather a quasi-aristocratic institution that has maintained itself for centuries by the constraints and buffers of promotion and tenure. You will have little to say about what goes on in the department. If you expect to make changes, you will be sorely disappointed.

SELECTING A GRADUATE SCHOOL

There is no foolproof way of assessing graduate programs, but there are a few steps you can take to ensure that you make the best possible decision. An appraisal of your interests and a systematic investigation of possible graduate programs will allow you to make an intelligent decision. Neither of these tasks should be taken lightly: you should budget about three months' time and begin your search for a graduate school in September, one year before you plan to enter.

Before deciding on a graduate program, you should undertake a thorough review of the geographic literature based upon your research interests, attempting to identify major trends of the past ten years, distinguish current trends, and evaluate possible future trends in geography. In addition, you should become familiar with the names of professors who appear to be the leaders in your field of interest and of those universities that offer ongoing research programs in your specialty. This review may seem boring and dull, particularly if you have read most of the material previously. However, the purpose of your investigation is not to evaluate the material on the basis of content, but on ideas, concepts, philosophy, points of view, and individual contributions. Three things should emerge from this investigation: (1) a better understanding of the field you have chosen for yourself, (2) a list of professors who share your geographic interests, and (3) a list of graduate schools that might be suited to your particular talents, abilities, and interests.

The next step is to reassure yourself that attending graduate school is something you really want to do. If so, then begin to look at the faculty members of your current department to determine whether any of them have graduated from the universities on your list. Make appointments to talk with them about the university; make sure the professor you speak with has recent experience with the program. A professor who graduated twenty years ago and has not maintained contact with his graduate department may know little of the current program because graduate programs change. If there are no faculty contacts available, prepare a letter of inquiry, and send it to the chairpersons of the departments you are interested in.

The following is a list of topics about which you might inquire:

1. Does the research program you have identified actually exist within the department? How long has it been a part of the department? How many faculty members actively take part in the program? Does the program interact with other research activities within the university? If so, with which ones?
2. How many degrees did the department grant during the past five years, and what were the thesis or dissertation topics? Which professors were most active as advisers? What kinds of jobs were found by the most recent department graduates?
3. How many faculty members does the department have, and how long have they been on the faculty? How many of the faculty have tenure? Does the faculty member with whom you are interested in working have tenure? Will he or she be on leave or on sabbatical in the near future—i.e., will he or she be there when you arrive?
4. What is the structure of the program? How many courses will be required for graduation? What types of courses are required? Given your current background in geography, will you have to take additional courses to make up deficiencies? Will you be able to take courses outside of the geography department?

5. What types of financial aid are available, including assistantships and fellowships? What is the deadline for admission? (Be sure to request all necessary application forms for both admission and financial aid.)

These are only a few of the topics about which you might inquire. The important point is to obtain as much information as possible that is relevant to your decision. You can answer some of these questions yourself by looking at the most recent issue of the *Guide to Departments of Geography in the United States and Canada* and the *Association of American Geographers Directory,* both published by the Association of American Geographers (AAG).

Once you have received replies to your letters, you should begin to evaluate each graduate school. Select from your list five that have demonstrated to you that they have the ongoing programs and faculty that best fit your interests. Reducing the list to five can be difficult, as many programs will appear similar, have strong faculty, and more than likely will have encouraged you to apply after receiving your initial letter of inquiry. These decisions can be critical because you may drop from your list a potential graduate program that fits your needs. So, make this cut carefully and discuss your decision with as many professors as possible.

Once you have reduced the list to your top five possibilities, the next step is to complete all the required forms—this task is not as simple as it might appear. First, it can be expensive, so find out ahead of time the application fee for each university. Second, give yourself plenty of time to complete the forms. Most graduate departments have simple fill-in-the-blank forms along with a requirement that you write an essay. Each graduate school will require a different essay, depending on the type of graduate student the department is interested in. This essay is probably the most crucial part of the application and should not be put off to the last minute. It should be written carefully, with no grammatical errors, no typographical errors, and no misspelled words. It should address the specific topics the department requests, not your general interests and background (unless specifically requested). Consider separately each graduate school you apply to; each will necessitate a distinct application. A general, all-purpose essay may gain you acceptance to the program, but not to financial aid.

Meet all deadlines, and ensure that your letters of reference have been sent to the department. Many students have failed to gain entrance to a graduate program because they missed deadlines, or because their references were slow in sending their letters. When requesting letters of reference, be sure that the person you are asking will write you a positive letter. You may think your abilities and talents are suitable for graduate school, but your references may think otherwise. Many students have failed to be accepted or have not received financial aid because their references have noted flaws and deficiencies. If you have come this far, make sure that your letters of reference contain no unwelcome surprises.

If you have evaluated your interests correctly, identified five graduate programs that appear to fit those interests, and completed the application forms carefully, you probably will begin to receive letters from the universities around the beginning of April. Your next concern is what to do if all accept you and offer you financial aid. There is no easy way around this dilemma. However, you can simplify the decision by carefully ranking each graduate program for which you submitted an application.

Have a clear and distinct procedure for ranking the programs, because that order is an important step in making a final decision. Base the ranking on your interests, how well the department matches your interests, its faculty, and its education potential. No two individuals necessarily will settle upon the same criteria to rank graduate programs. Once you have ranked the programs, making your selection will be easy if your Number One department accepts you and offers financial assistance. However, the world is not always as simple as ranking and accepting. Your top-ranked department may accept you but decline to offer financial aid; your top-ranked department is probably many other students' Number One choice also. Remember, each department may accept you with financial aid, accept you without financial aid, or reject your application. You probably will receive various combinations of acceptance, rejection, and financial aid from among your top five choices.

At this point you should accept the highest-ranked program that accepted you and offered financial aid, and not the one that offered the largest financial aid package; remember why you ranked the programs. After you have made your decision, promptly inform all of the universities to which you submitted applications of your decision. A prompt and courteous letter will allow departments to make adjustments in their offers to other students.

WHAT TO EXPECT IN GRADUATE SCHOOL

Now that you have been accepted to graduate school, what should you expect? Each graduate department is different, so no matter which you choose, you can expect that it will differ significantly from your current department. Probably the most apparent difference is that professors pay more attention to graduate students; you suddenly become someone the faculty members know on a first-name basis. In addition, learning in graduate school is quite different from the undergraduate experience: There is little that is true or false. What was once presented as a fact is now seen as a point of debate. The transition from undergraduate to graduate student is a transition from absolute to relative, an adventure into the realm of ideas, concepts, methodology, and philosophy. You will find yourself challenged by new ideas and different points of view.

As a beginning graduate student, you may want to know what a department expects from its graduate students. Again, there is no single, specific answer for all graduate departments, but a few general comments may be helpful. Most graduate departments expect their students to be responsible, have integrity, and be highly motivated to study geography. Students are expected to complete their assigned work without complaints or excuses. In terms of research, the beginning student is usually expected to demonstrate an understanding of the subject and, above all, the ability to learn. Advanced graduate students are expected to demonstrate a command of their subject as well as the ability to conceive, organize, and conduct a research project without a great deal of handholding from the faculty. For both the beginning and advanced graduate student, the most important skill is writing. The ability to write clearly and effectively is half the battle of graduate school. You might be the brightest student in graduate school, but if you cannot communicate your ideas, no one will know.

A question most new graduate students ask is, "What is a seminar?" Describing a seminar, particularly a good one, is very difficult. In general, a graduate seminar differs from an undergraduate class in that the student is expected to be an active participant; the professor should provide the setting and act as a guide through the seminar topic. However, every professor conducts a seminar differently; no two are exactly alike because of the intangible elements that go into creating one.

It is much easier to identify the characteristics of a bad seminar, the one you should avoid if at all possible. It will usually begin with little or no focus on a specific topic and will be poorly organized. The course outline will appear to be inconsistent, and the reading list, if there is one, will contain few readings, many of which are not current. What the students are expected to do in the seminar won't be identified clearly. The professor will place little emphasis on concepts, will downplay the importance of developing research methodologies, and won't stress the need for critical thinking. Almost no assistance will be given to help students develop research topics. During seminars the professor will lecture a great deal, allow very little student discussion, provide only one idea (usually his or hers), and will rarely stay around after the seminar to answer questions. Assigned readings won't be followed routinely by a lively discussion in the seminar. In some cases, professors have been known to give multiple-choice exams at the end of the semester to test graduate students. If you find yourself in such a seminar, try your best to find another. Above all, graduate seminars should be challenging, exciting, and stimulating.

The success of a graduate seminar, however, depends largely on students. Students who have completed assignments before class begins, who are active in class discussions, and who are willing to listen and accept differing ideas and interpretations, rarely will leave a seminar without having learned a great deal.

Of course, your workload will increase tremendously in graduate school, and you will read more and write significantly more than you might expect. For example, a seminar report is not the usual undergraduate term paper quickly thrown together a few days before it is due; ideally, it is a thoughtful, well-researched paper focused on a specific topic, which usually runs from thirty to fifty pages with maps. Depending on the number of seminars taken, you can expect to prepare from two to four seminar reports each semester.

If you arrive with a B.S. or B.A., you will be expected to perform at a certain level; if you have already obtained an M.S. or M.A., you will be expected to perform at a higher level. Most graduate schools expect beginning graduate students to have a basic background in geography; if you do not, expect to make up deficiencies in your previous undergraduate or graduate training by enrolling in additional courses. These courses are not remedial, but are intended to provide you with the necessary breadth and background in geography, which, in turn, will allow you to get the most out of your stay in graduate school.

The transition to graduate school can be difficult, and it is normal to feel baffled and discouraged at times. It is also normal for graduate students to complain about the workload, the faculty, the library, and almost everything—in fact, at most universities this is expected. However, complaining can become a habit and thus make graduate school a frustrating experience. If you are a complainer by nature, you can be assured that graduate school will provide you with much to complain about. Complaining too much, however, is self-defeating and surely will stifle your education. There will be ample opportunity to complain after you have graduated, especially if you choose to become a university professor!

Graduate school can be a great deal of fun, particularly if you have an unflagging interest in geography. The interaction between students and faculty, the seminar readings, the lengthy discussions of different theories and concepts, the discovery of new ideas and different interpretations, and, finally, the challenge to learn how to learn are what graduate school is all about. Your ultimate success will depend, to some extent, on your work ethic and almost certainly on your willingness to embrace new ideas. Ideally, graduate school is a combination of students and professors working together to create a setting for scholarly and scientific imagination.

CONCLUSION

This essay has attempted to identify a procedure for reviewing graduate schools and has briefly described what a new graduate student should expect. No two graduate schools are exactly the same; therefore, the suggestions presented here should be taken not as absolute but should be weighted heavily with common sense. The transition from undergraduate to graduate student can be a trying experience; a graduate education is demanding, and the course work is exacting. While each student adjusts differently to graduate school, those who have the ability to accept and seek new ideas, and who possess a willingness to work hard and have fun while doing it, are usually the most successful.

Relations Between Geography and Other Disciplines

John E. Chappell, Jr.

More than most other fields of study, geography is a *correlative* discipline: the knowledge it produces and teaches is built in large measure out of selection and combination of results from other disciplines, such as geology, meteorology, history, anthropology, and economics. Even those concepts and theories used to produce distinctively geographical knowledge, such as Alfred Weber's theory of industrial location, are often taken from some other field. The chief aim of this chapter is to facilitate the way in which geographers approach and utilize knowledge from some other disciplines, especially in certain newly explored areas of social theory and philosophy. The treatment will be straightforward and elementary, taking care to correct a few shortcomings that have persisted in the literature.

Of many possible examples and case studies, those I have chosen are related primarily to my own special interests, in philosophy, general intellectual history, history of science, physics, climatic change study, and Soviet area studies. As this list implies, I have studied mostly in other disciplines, although I have taught and published mainly as a geographer.

Building relationships with other fields has become a crucial matter for geography, in view of recent loss of disciplinary status in several major universities. Geographers need to understand and utilize work in other disciplines not only to advance their own perspectives, but also to demonstrate to scholars in other disciplines—who collectively pass judgment on such matters as whether or not a field merits disciplinary status—that their own scholarship does in fact possess the degree of integrity, maturity, validity, and significance that entitles it to such a status.

Elsewhere I have suggested that one of the surest means of reestablishing respect for geography would be to bring about a revitalization and more convincing rendering of the theme of influences of natural environment on behavior and culture (Chappell 1975, 1981). This strong conviction will be expressed here as well, with the aim not of stirring up controversy, so much as of simply strengthening and clarifying my general message.

At the outset it seems wise to pause for consideration of just where geography ends and other disciplines begin: in other words, to ask again that much debated question of how to define "geography." In such matters I believe that the more inclusive definitions, which aim for tolerance and for a broad consensus, tend to be the best.

Geography has already been hindered for many years by narrow definitions. Beginning in the 1920s, Carl Sauer often accused William Morris Davis of trying to limit geography to searching for ways in which human culture is influenced by physical environment (Sauer 1963). The new consensus as to the proper concern of geography was soon labeled by Richard Hartshorne (1939) as "areal differentiation." But meanwhile a new kind of narrowness began to be enforced, resulting in half a century of neglect of natural-environmental influences. Yet it has never been proven that such influences, sensed to be real by most ordinary people, do not operate in human life (Rostlund 1956). Surely the matter could be investigated again without naïvely duplicating errors in previous approaches.

Ignoring half of the total two-way ecologic relationship has been accompanied by an overemphasis, by quantifiers and humanists alike, on the concept of "space" as the central concern of the field. At times, causal powers seem to be attributed to the dimension itself, as if, quite contrary to known laws of nature, it exerted a physical force. But *every* scholar deals with space as a dimension; it infuses all reality and cannot be ignored. Geographers clearly have no *exclusive* rights to study it; at most, their role might be to emphasize it more than do most other disciplines.

Yet emphasizing space, or spatial patterns, or spatial interaction, or society in space does not seem to have done much lately for the prestige of the discipline. I suggest that this is not because the spatial approach is fundamentally wrong, but because all by itself, it leaves too much out, making geography appear to constitute something less than a complete discipline.

One thing too often left out is the dimension of *time*. Since Sauer and his school of historical-cultural geographers argued more or less successfully against Hartshorne and others that the temporal factor should have an important place in the field, the same old suspicion that time should be left to the historians has cropped up again—even reinforced, in some discussions, by a feeling that "time" is some sort of hostile concept, challenging the "space" of geography for supremacy. But *both* dimensions are essential to *all* kinds of study. For example, the same migration or invasion that interests historians might also be studied as a topic in geography, but not without allowing for time, as well as space, in which the process can occur—just as historians must consider space as well as time in their own approach.

Some would limit formal geographic study to human society, leaving subjects like geomorphology and meteorology to other disciplines, on the grounds that human beings are far more important than nonhuman nature. Yet engineers, chemists, and many other scholars who rarely if ever study human beings in their professional roles may be convinced of the importance of human life and values, at the same time they realize that these ends cannot be served adequately without studying the nonhuman part of nature. Without such study, we would not even have calendars, by which we know when to plant the crops that sustain our lives.

Once the natural environment receives attention, relationships between it and its human occupants routinely come into view. Yet an overemphasis on space and culture, and an underemphasis on natural science and natural environment, have obscured the full range of these ecological relationships, despite some continuing attention to certain ecological themes. Over the centuries, many other disciplines also have explored human-environment relationships (Glacken 1967), but no other field would appear to have as clear a mandate to do so, as one with "ge-" (earth) in its name.

The best answer, I believe, is to suggest *the surface of the earth* as the substantive object of study for geographers. This term would be meant to include the entire biosphere, from deep in the earth's crust to the top of the atmosphere, and all life and culture within it. It happens that differentiating between the many parts of this broad realm, and studying relations between them, requires spatial awareness and spatial concepts. But any part of the biosphere can also be divided into ecological components, the study of which requires a different set of concepts. One would rarely go wrong in categorizing a study as truly geographical when it involves attention to *both* spatial *and* ecological relationships—i.e., to both *situation* and *site*. The same study would assume much more significance, if it also paid attention to how those relationships developed over time.

This approach may appear to leave out certain efforts, such as mapping of phenomena without exploring any ecological or temporal relations, or study of structures in the "built environment," with little attention to either situation or site. Such work of limited scope, as well as the many minor specialties whose proliferation has quite understandably been accused of obliterating the focus of the discipline, need not be disqualified out of hand. Instead, it might be wiser simply to keep pointing out what rich and meaningful geographical scholarship is all about, and leave it up to the good sense of individual geographers to guard their own respectability by maintaining the standards. And after all, an apparently trivial project might involve a larger vision, toward which it is only a necessary first step.

PROBLEM-SOLVING AND INSULARITY

In any case, the area of investigation just defined remains so vast, that some means of narrowing down its scope seems called for. In this regard I strongly recommend the *problem-solving* approach put forth by Edward Ackerman (1963) and already illustrated in the careers of numerous geographers. If, from the nearly endless range of subject matter on the surface of the earth, one chooses material closely related to a pressing human problem, the issue of what is or is not properly geographical tends to fade into the background. It is difficult to take such an issue too seriously when embroiled in such urgent matters as the problems of war and peace, justice and human rights, hunger and overpopulation, or the degradation of the environment. Of course smaller, less urgent issues also require treatment. We cannot pay all our attention only to those matters currently in the forefront of consciousness, or we would be laying inadequate groundwork to deal with unpredictable problems that are likely to crop up later.

Ackerman also emphasized the importance of studying ecological relationships within the earth system, where a great many of the crucial problems of the near future are likely to be found. The fact that these and most other urgent challenges to our minds are inherently interdisciplinary in nature provides a useful check against excessive disciplinary insularity, and against the tendency to remain smug and comfortable within one's own special intellectual territory. The real world, after all, is not divided neatly into such categories.

In choosing a worthwhile problem for geographic research, it may be wiser to find one that *is* being treated by a number of workers in other disciplines, than one which is *not*. That way, one is more likely to find a problem of considerable significance. Even if it yields no identifiably "geographical" insight, this kind of effort seems more likely to improve the image of geography, than would the procedure of selecting and studying a problem in terms of its amenability to treatment as an exercise in spatial relations, and then declaring, in effect: "See what we geographers have come up with, that you spatially-illiterate other disciplines did not find!" Genuine respect comes not so much from special status as from willingness to take on difficult and necessary tasks with a minimum of self-importance.

If a geographer does happen to be the one who makes some crucial advance in solving an important problem, it seems more likely to happen because of sensitivity to important needs, and an ingrained habit of tirelessly searching out relevant facts and concepts, than because of any unique spatial orientation. In any case, beyond all that has been said so far, probably the best reasons why geographers should put their talents and energies to solving the world's crucial problems are simply that all other disciplines together cannot provide enough, and that in some cases—such as in the current drought and famine crisis in East Africa—human needs cry out so loudly as to imply a moral obligation to help.

Another kind of insularity afflicts geographic scholarship, a most inappropriate one for a discipline supposed to deal with the varied panorama of life on the earth: namely, a very limited knowledge of foreign cultures. To attain thorough knowledge of a foreign culture of course requires study in other disciplines, such as history, language, and literature. But the replacement of foreign languages by quantitative skills in many doctoral programs, declining popularity of regional courses, and perhaps even an infusion into the profession of the cultural insularity that has helped to produce the current appalling state of geographic illiteracy in the general population, have helped to work against such cross-disciplinary labor. Consequently it seems at times that multidisciplinary regional studies programs achieve more thoroughness in understanding geographic regions than does geography. All it takes to repair such a deficiency, really, is dedication and hard work.

AWARENESS, ACCURACY, AND CONTEXT

Treatment of extra-geographical subject matter begins with an *awareness* of what it consists of, and the cleverness to select from it the most appropriate portions.

Frequently such appropriate material is indeed selected, but occasionally, important work of potential relevance to geography is too long overlooked. Let me offer an example related to the 1988 meeting of the Association of American Geographers in Phoenix.

At this gathering, several sessions were held on the topic of climatic change. One of them paid appropriate attention to the fine work by Harold Fritts and his Laboratory of Tree-Ring Research at the University of Arizona. But absent from the program was any attention to an even more closely local, more recently controversial, and perhaps more theoretically interesting body of scholarship on climatic change theory, which has emanated for well over a decade from the Phoenix offices of the U.S. Department of Agriculture only a few miles from the conference hotels: the numerous publications of meteorologist Sherwood Idso. Lack of interest in Idso's work among geographers may have something to do with the fact that it often implies we should reduce the recently fashionable emphasis placed upon the causal influence of human beings on their environment. For instance, Idso and others, including Benjamin Herman at the University of Arizona, began early in the 1970s to show that the kind and location of dust that humans place in the atmosphere probably has more of a warming than a cooling effect, by adding to the greenhouse effect (Idso 1981). As a result, decreasing attention was paid to suggestions that worldwide atmospheric cooling between the 1940s and 1970s might have been caused by human-produced dust, leaving little besides astronomical causes like solar cycles as the most likely candidates. Idso has also been a leader in the unpopular struggle to revise significantly downward various widely publicized, alarmist estimates of a harmful warming effect from increasing human-produced carbon dioxide (1985).

Once appropriate extra-geographic material has been selected, it is of course vitally important to achieve *accuracy* in understanding, to learn definitions, facts, and concepts with care and precision. This is not always done. One often gets the impression that once a geographer has learned more about a new nongeographical topic than his colleagues are likely to know, he or she rushes ahead in an effort to impress them, without adequate command of the material. Incorrect interpretations may be made, and we may not obtain the elucidating simplicity in dealing with basic concepts that real experts seem best able to achieve. Before explaining to geographers some new concept from another discipline (as opposed to merely mentioning it, as an intriguing possibility for study), one might be wise to ask oneself if this explanation would satisfy a member of that discipline.

Ideas from the past, geographical and otherwise, are often dismissed with some variety of the one-line put-down, employing such phrases as "failed to demonstrate," "uncritical acceptance," or "mere assertion." But if such a put-down is unaccompanied by careful refutation, built on accurate knowledge of existing texts, or at least by citation of such a refutation, it may amount only to "mere assertion" itself, "failing to demonstrate" much other than bias in favor of the most recent intellectual fashions.

It is often of crucial importance to place basic facts and currents of thought in their proper *context*. Otherwise, one may wade in with preconceived expectations and emerge with isolated passages that misconstrue the total picture. Because of human fallibility, and a natural propensity to grow and change over time, this is especially easy to do in the case of passages from an author who produced great quantities of material. Proper interpretation may require appreciating various subtle nuances, and consulting additional sources, in order to place the passage in its wider context.

Several examples of problems in achieving accuracy in understanding, and of appreciating full and proper context, will appear in the next three sections, which are devoted to certain key thinkers and ideas from social science, natural science, and philosophy. These will be presented in terms of their potential usefulness to the human geographer.

MARX, WEBER, AND SPENCER

Gone are the days when geographers were considered well-informed about theoretical matters if they had read a reasonable segment of Hartshorne (1939), and a few key articles by Sauer, and could outline basic trends in geographic thought over about the past two centuries. In view of recent strenuous efforts to achieve a wider interdisciplinary perspective, most of them made by British Commonwealth geographers, one must now be familiar with numerous leading trends in philosophy and social theory that were scarcely recognized or believed to be relevant to geography as little as twenty years ago. (One often gets the impression that geographers following these ambitious new pathways are leaving the earth behind and trying to work primarily in some other discipline. But we should not invoke disciplinary narrowness too hastily; the geographic relevance may soon follow, and it is the general importance to humanity that matters most in any event.)

From among the thinkers being studied in this search for expanded horizons I will mention here only three, whose names and ideas are of broad import throughout the social sciences. Naturally, many other names and ideas are also worth learning about; this review is only an introduction.

Karl Marx is of course not a new and unfamiliar name. And it is not only since the period of the Vietnam War that an ambitious minority of geographers, many of them associated with the journal *Antipode,* have considered study and discussion of his ideas an integral part of their disciplinary task. Specialists in the Soviet, East European, and Chinese regions had long been studying Marxism as part of the culture of the nations therein, concentrating primarily on its applications and transformations in the real world.

But what may be new is the possibility that, before long, human geographers in general will consider it their obligation to study Marxism, as do many scholars in other social science disciplines, regardless of their degree of sympathy with it. It will seem obvious to some and vexing to others to say this, but human geographers need to know something about Marx because they belong not only to geography but also to social science, and Marx has been clearly the most influential social scientist in modern times, in a scholarly as well as a real world political sense. In most large and influential universities, he is taken very seriously by scholars who do not belong to the political far left. Among these, I have heard sociologist Daniel Bell argue that Marxism is "still our best tool for analyzing bourgeois capitalist society," even though it falls short in other ways, such as in analyzing socialist societies–none existed when Marx and Engels lived–and in coping with the problem of resource scarcity (Bell 1975).

Marx also erred in predicting that his brand of socialism would develop first in those nations that first became capitalistic. In fact, such nations eliminated many (but not all) of the harsh injustices of early capitalism that motivated Marx's crusade, through legislation and gradual change; and Marxist revolutions succeeded first in less fully developed lands. Such differences in societal evolution in different parts of the earth pose a problem with obvious geographic dimensions, and it remains to be solved.

There is no shortage, in such sources as *Antipode* and the works of David Harvey, of practically oriented elaboration of Marxist concepts and themes, such as the labor theory of value or dialectical materialism, or Lenin's modifications such as his analysis of imperialism. Less partisan grounding in the basics can be obtained from Lichtheim's standard survey of Marxism (1961). Often discussed by geographers is post-1917 Marxist thought in the West, involving Lukács, Adorno, Habermas, and others; for such ideas one might explore two pithy books by Perry Anderson (1976, 1984) and the large collection edited by Arato and Gebhardt (1987).

Next to Marx, the most discussed social scientist of the past century or so has been Max Weber, the more prominent brother of Alfred Weber. Max Weber was motivated largely by an antipathy for Marxist materialism and scientism. If Marx argued that religious beliefs are rooted in economic facts, Weber countered that economic facts are often rooted in religious beliefs. If Marx emphasized the role of the group and of impersonal historical forces, Weber preferred to stress the role of the individual and of willful human behavior. In contrast to Marx's simplified view of class conflict, Weber built a more intricate picture of social and political groups, with emphasis on types of authority (including "charismatic," a term he popularized) and the complexities of bureaucracy.

But Weber was far more than an ideological disputant, and he tried strenuously to avoid extreme positions. Although the importance he attached to personal values and individual behavior has led to his being viewed by some later thinkers as a sort of fountainhead of subjectivist thought, he did respect scientific goals and went beyond a merely idiographic approach to seek "ideal types," opposing patternless relativism and particularism (Weber 1947).

Max Weber's thought has been cited widely by geographers interested in promoting humanistic approaches. But they generally ignore the one publication that did far more than anything else he wrote to make him a center of controversy and respect: *The Protestant Ethic and the Spirit of Capitalism* (1930). Here Weber argues that Calvinist asceticism provided values and incentives that led to early capitalist accumulation and organization, thus doing a great deal to define the nature of the modern world. This book induced a flood of scholarship attempting to explain the origin of capitalism, most of it disagreeing with Weber and finding the chief causal influence in the economic realm (Tawney 1946; Samuelsson 1961). One would suppose that geographers interested in capitalism, either as opponents or supporters, or in the theme of "postmodernism"—which seems to require consideration of the nature and origin of the modern—could find some value in this controversy.

Another sociologist, about a generation earlier than Weber, provided an even more pronounced alternative to Marxism: Herbert Spencer. Unlike Weber, Spencer is rarely admired today, but his views need to be known, if only to help define the alternatives to them by providing larger context and sharper contrast. In 1850, Spencer began what became a huge and very influential project, with a book extending evolutionary concepts into the social realm. He declared that the ruthless struggle for existence that characterizes raw nature occurs also in human society and should even be encouraged to operate there, by avoiding government regulation of the economy. Thus, he thought, the strongest would flourish and multiply, improving the race as a whole. He interpreted suffering and even death among the losers in this struggle as part of a generally beneficial process (Spencer 1892).

Although he introduced it before Darwin's biological theory of natural selection appeared, Spencer's social theory, along with the similar views of John Fiske, William Graham Sumner, and others, later became known as *social Darwinism*. Around 1900, this doctrine was being soundly refuted by numerous scholars including theologians, sociologists, and the Russian geographer Peter Kropotkin (Hofstadter 1959); but somewhat less inhumane variations on it keep reappearing in right-wing political programs, usually under some other name and guise, such as an alleged defense of "free enterprise."

Social Darwinism has been mentioned rather often lately in geographical literature, but not primarily in relation to Marxism. Often ignoring the standard definition of the term, and perhaps wrongly assuming it refers to *any* sort of application of Darwinian ideas to social science, several geographers since the time of Leighly (1955) have interpreted social Darwinism as closely related to environmental determinism, in an ill-conceived campaign to discredit the latter. In actual practice by its proponents and leading interpreters, social Darwinism refers only to relations between humans, and not to human-environment interactions (Chappell 1988).

EVOLUTION AND THE NATURAL SCIENCES

Considering the widespread inclination to find analogies to evolutionary concepts in social processes, it seems wise for social scientists to have a basic familiarity with the theory of evolution. I have long believed that a course in history of geographic thought is incomplete without a study of evolutionary theory, in which factors of both site and situation play prominent roles. (A fine book for such purposes is Eiseley 1958.)

The extremes of social Darwinism served to warn many scholars against a too-facile application of evolutionary concepts in the social realm. A broad reaction against all "organic analogies," including evolutionary and environmental-determinist ideas in general, soon occurred in the social sciences. Partly through the influence of anthropologist Franz Boas, this trend was strongly reflected in geography by the work of Sauer and Leighly. Yet overall, the reaction was too extreme, and ultimately it left geographers ill-prepared to make important contributions to the new ecological awareness that began to overwhelm both the scholars and the general public about 1970. After all, one can scarcely take ecology seriously while at the same time holding on to the fiction that human life goes on without significant involvement with biological or environmental factors. Yet this fiction keeps on being promoted, as in Gregory's complaint about "misleading theoretical analogies which project from one domain on to the other," among which he includes systems theory, an important tool of ecological analysis (1978, 171).

Of course human life proceeds on a level higher than that of the animal world in general. Yet it is also true that, whatever else they are, humans do also belong to the animal kingdom, and to both the organic and the physical realms in general. If our biological nature did not matter, we would not need water or food to survive, and if the physical environment did not affect us, we would not freeze when unprotected in subzero temperatures. Such types of influences also operate in many less obvious ways.

A widely debated new effort to use biological factors to explain social phenomena is E. O. Wilson's *sociobiology* (1978). Popular with some biologists and others, this version of genetic determinism also has many opponents, who place greater emphasis on causes in training and environment.

Evolutionary theory gives strong support to the notion of the causal power of natural environment. Given the occurrence of variations among individual offspring, the selective action of the natural environment leads to evolutionary development, whenever that environment changes. (If it does not change, species remain fixed.) *Natural selection* is clearly an environmental-determinist process, favoring those organisms best adapted in the struggle for survival. But competition is not the only factor; as Kropotkin showed in refuting social Darwinism, *cooperation* among members of one species—from ants and bees to wolves and bears—helps the entire group survive. Using an organic analogy much different from Spencer's, Kropotkin then argued that cooperation among human beings yields a similar advantage (1972).

An evolutionary process that has often been invoked—by Spencer, by many of his critics, and even by Darwin himself as a supplement to natural selection—but which nearly all biologists today do *not* believe operates in nature (Gould 1980), is *Lamarckism*. According to Lamarck, a pre-Darwinian evolutionist, the driving force of evolution is the use and disuse of bodily organs in the process of striving to meet environmental challenges (Ingold 1986, 226). Individual behavior is crucial, and bodily characteristics thereby acquired are inherited. (Later, the concept of direct induction of adaptive—as opposed to random—genetic change by environment, known as "Geoffroism," became incorporated into the term "Lamarckism.") Even if Lamarckian processes do occur, they could only supplement, and not replace, the enormous influence of selection by natural environment.

Current evolutionary theory involves much more than just the causal influence of environment—e.g., the concept of *chance,* which is often held responsible for the variations on which natural selection operates. By focusing on this factor, pragmatist philosopher William James drew from Darwinism not a vision of mechanistic influences and an inevitable process of struggle, but a message of open possibilities (Chappell 1988).

Social scientists at times draw on concepts in *modern physics,* and especially quantum physics (which deals with very small particles), for justification of the idea that the universe in general runs not by mechanistic laws, but according to probabilistic or stochastic processes. In other words, if even the most exact of sciences must retreat from belief in mechanistic laws, why should not social science do likewise? But in dealing with modern (i.e., twentieth-century) physics, one must be very careful to distinguish proven fact from sheer speculation. Speculation with no empirical basis is often found in articles by physicists in *Scientific American* that deal with such imaginary concepts as backward-flowing time or eleven-dimensional "space-time." In his presidential address to the Geological Society of America, energy expert M. King Hubbert accused physicists of a careless attitude toward fundamental definitions and principles, leading to many errors in their own basic handbook (1963). Dissident theoreticians have found even more serious problems than these in modern physical theory (Chappell 1979).

As for replacing traditional mechanistic concepts by probabilistic ones, it should be noted that many scholars interpret the word "chance" as primarily a term to cover what is not yet understood. One does not have to give up belief in mechanistic causation in order to speak of chance occurrences. Surely Darwin believed that the world is governed by deterministic processes, even though they might often elude our understanding. Some interpreters also approach quantum physics in this spirit; they claim that what has not been, and perhaps cannot be, measured at the quantum level may not represent an uncertainty inherent in nature, but only the inability of the investigator to discover all that might be known.

Recent interdisciplinary work on "chaotic" phenomena, much of it arising from concerns with long-range weather forecasting, may be interpreted as casting doubt on the idea that change or randomness is inherent in nature. Deterministic nonlinear functions can often be shown to be responsible for what earlier were considered examples of mere chaos (Gleick 1987).

There is of course much more to physics than the "modern" variety. Geographer and geomorphologist Michael Woldenberg, in his study of branching fluvial networks, has demonstrated that where statistical methods fall short and where quantum physics does not apply, traditional *Newtonian physics* may provide long-sought answers. In this work, which applies to blood vessels as well as streams, and thus crosses over into biology, Woldenberg has used an "equilibrium of forces" model (1986). This is the same model, utilized in a different way, that Alfred Weber employed to build his theory of industrial location; Woldenberg knew the points and reasoned from them to the rules, while Weber started with the rules and used them to determine the points.

PHILOSOPHICAL CONCEPTS AND TRENDS

Philosophical terms and concepts intrude deeply into social science theory, and at least a few of them need to be understood in geography. For example, it is hard to read any theoretical discussion these days that does not refer to *positivism*—usually as an object of disapproval. But "positivism" is rarely defined. Just what does it mean? In many contexts, it appears to be a near-synonym for "science," or "the scientific point of view"; or it may seem to refer to any claim of being "positive" that one's results are correct. At any rate, it is clear that numerous critics of positivism have adopted a highly relativist or subjectivist point of view, doubtful of the very idea that there are sure and correct facts out there to be discovered. The implication is that to avoid the evils of positivism, we had better not be too "positive" about what exists in the environment beyond our perceptions—not to mention what influence it might have on our lives.

But like social Darwinism, the term positivism has had a narrower range of meaning in actual usage by its protagonists, than recent usage in geography would suggest. That range does not by any means include the whole of scientific activity. Positivism was the term applied by Auguste Comte to his new sociology in the early nineteenth century, and by it he meant that he aimed at *sure* knowledge, and at the avoidance of mere speculative metaphysics. But Comte—like Spencer, who is also at times called positivistic—departed widely from this ideal, by introducing his own speculative theory about societal development.

The essential meaning of positivism is better exemplified by a version of it known as "critical positivism" or "empiriocriticism," practiced by various natural scientists and philosophers near the end of the nineteenth century (Abbagnano 1967). The most prominent of these was Ernst Mach. Although as a physicist, Mach could hardly avoid theory, he was very skeptical about any theory not firmly rooted in observational fact; he even doubted the atomic theory of matter, in an era when atoms could not be seen. Mach exerted great influence on subsequent developments in physics, as well as on the *logical positivism* that soon emerged from the Vienna Circle philosophers.

Logical positivism placed more emphasis than Mach did on the spinning out of theories, by means of careful use of language and logic. Accordingly, quantitative geography, fashioned largely in the spirit of this latest form of positivism, tends to emphasize the building of theory more than the gathering of facts. Thereby the original, most fundamental emphasis of positivism, which is on believing only in well-confirmed facts, is somewhat obscured. It is also obscured by the lack of mention of Machian positivism in some wide-ranging and generally very useful surveys of philosophy and social theory written by geographers (Gregory 1978; Jackson and Smith 1984).

Other philosophical terms that need to be understood by geographers include, for example, *metaphysics,* the near-antonym of positivism. One engaged in metaphysics is aiming to discover ultimate truth. Because this effort usually goes far beyond empirical facts, it usually involves considerable speculation; but perhaps the adjective "speculative" ought to be attached to *metaphysics* to specify such a circumstance, since metaphysics *can* be carried out with rigorous logic, so that its conclusions about ultimate reality are tightly connected with empirical facts.

Phenomenology has been widely discussed in geography as an alternative to positivism, to provide philosophical underpinnings for environmental perception studies and humanistic geography. It dwells on personal sense perceptions as the source of knowledge and strives to avoid biases implicit in traditional theory and terminology. Geographers who deal with phenomenology often fail to notice that it began as an effort to do a better job at the same task positivists aimed to do: being certain about one's facts and avoiding unfounded speculation. Edmund Husserl, who initiated modern phenomenology around the year 1900, even wrote that "*we*. . .are the genuine positivists" (1931, 86).

Phenomenologists emphasize the creative "intentionality" involved in the act of perception, and seek to be more thorough and careful at noticing every nuance in perception, than the positivist who allegedly rushes in looking for something to measure and ignores much else. But they are by no means necessarily subjectivist in their approach to knowledge, as geographic interpreters often seem too quick to assume (Relph 1970; Jackson and Smith 1984). Most phenomenologists, including Husserl, definitely *affirm* the idea that there is objective or "intersubjective" truth outside of our perceiving selves, which different observers can agree upon (Spiegelberg 1965, vol. 2, 668). In fact, one of the main purposes of Husserl's phenomenological method is to clear away merely subjective elements in order to focus on essences, or "invariants." (In his final writings, however, perhaps influenced by the work of his existentially oriented student Martin Heidegger, Husserl may have retreated partially from this affirmation.)

Some geographers have shown particular interest in the socially oriented phenomenologist Alfred Schütz, who taught in the United States around mid-century. Schütz combined the sociology of Max Weber with Husserl's concept of the "life-world" of ordinary experience (Spiegelberg 1965, vol. 2, 631).

Existentialism also concentrates more than Husserl did on real life individual experience and emphasizes the anxiety and forlornness accompanying human freedom. Jean-Paul Sartre, its most typical representative, claimed that ongoing human "existence" precedes human "essence," which is defined by choice and experience. Thereby he denied a fixed human nature and also cast doubt on whether there is an objective reality accessible to science.

A recent and extreme development of this subjectivist trend is the *deconstructionism* of various French writers, including Jacques Derrida. Here, not only objectivity, but even the *foundationalism*—belief in some secure foundation for truth, be it objective or subjective—that had accompanied almost all previous philosophy, have been decisively abandoned (Megill 1985). Derrida's nihilistic thought is also called "post-structuralist," in that it opposes the earlier *structuralism* of anthropology and linguistics, which assigned to abstract patterns such as language structures a controlling influence over particular facts. Derrida not only abandons the search for scientific objectivity and truth, but he also ridicules grammatical rules, and thinks the truest language is that most independent of human purposes (Harland 1987). Evidently, in philosophy as well as physics, one must be wary of adhering to a trend merely because it is the latest thing.

Deconstructionism is sometimes claimed to offer a path into a hypothetical postmodern world, which would be free from the mechanistic and insensitive features of modernism. Fortunately, several theologians, philosophers, and scientists have also been promoting a more constructive form of *postmodernism* (Griffin 1988). Such scholars endorse *ecology* as a typical postmodern science—more subtle and complex than much that preceded it, but not at all denying the validity of scientific effort in general.

Some geographers have recently become interested in *realism,* which in its philosophical sense refers to belief in the existence of a real world beyond perception, which the perceiving person has not created (Wild 1948). This common sense viewpoint was defended by G. E. Moore and others early in the twentieth century, and was implicit in both positivism and Husserl's phenomenology (Chisholm 1960); it has since been challenged by the subjectivist trends just described.

Beyond the considerable scientific implications of positivism, *philosophy of science* has entered geography in recent years mainly in the form of Thomas Kuhn's widely influential idea that scientific revolutions occur by means of a wholesale shift in *paradigm*, which may at times result from aesthetic, psychological, or other nonscientific motivations. A paradigm is a broad theoretical outlook that defines not only how facts are interpreted, but also how research problems are selected and dealt with (Kuhn 1970). Although showing how subjective factors can be influential in science, Kuhn does not, as some have implied, abandon belief in objective scientific progress.

Competing philosophical views may both have some validity in the same area. For example, a positivisitic approach may yield useful statistical findings about human problems, while at the same time, as phenomenology suggests, the same data may reveal other layers that are not susceptible to precise measurement or reducible to simpler elements. In the case of polarized viewpoints, it is often wise to take Aristotle's advice to seek the mean between two unsound extremes.

In philosophy (more than in most areas), the accurate essence of an idea may prove very elusive in the limited time a geographer usually has to read primary sources in other fields. It is thus important to supplement such reading with reliable secondary works. In turn, referring to more than one of these may help in judging their reliability.

Besides general encyclopedias, certain specialized reference sets can provide sound information. The *Encyclopedia of Philosophy* is generally very reliable and useful, as is its counterpart for social theory, the *International Encyclopedia of the Social Sciences*. Works in allied fields also can be valuable: *The Dictionary of Scientific Biography* and Mircea Eliade's new *Encyclopedia of Religion* provide considerable information on those philosophers who were also important in natural science and religion, respectively. Considerable help can also be obtained from collections of primary sources. Of these, the six-volume New American Library series, with lengthy commentaries by the editors, is especially helpful; it includes a fine survey of early twentieth-century philosophy edited by Morton White (1955).

Note that there is also a subdiscipline known as "intellectual history," which encompasses ideas in many areas beyond philosophy *per se,* in the arts, social sciences, politics, and so forth. It has its own distinctive reference works, such as the *Dictionary of the History of Ideas,* collections of readings (Baumer 1964), textbooks, and monographs.

Philosophers do not always deal only with abstruse theory. Some have shown a penchant for getting involved in the real problems of the day—even those problems connected with the study of the earth. One such trend in philosophy is the *deep ecology* movement, which seeks a new ethic in human use of the land. At times it seems quite impractically idealistic, but it does offer an interesting challenge in an important area (see Duvall and Sessions 1985; and such journals as *Environmental Ethics, Inquiry,* and *The Trumpeter*.) Deep ecology carries on with themes common in the debate initiated by historian of technology Lynn White, Jr., when in 1967 he blamed Christian theology for inspiring a destructive attitude toward the environment (Barbour 1973).

Finally, a small specialty called *the philosophy of agriculture*, is also concerned with ethics and values, particularly in connection with the application of agricultural science and technology. It deals with injustices not only in North America (see Dundon 1986; Friedland and Kappel 1979), but also in the Third World in relation to the agricultural imperialism discussed in the classic study by Lappé and Collins

(1977). Scholars from several disciplines write on these themes in the journal *Agriculture and Human Values,* edited by philosophers at the University of Florida. Such concern over practical matters on the part of a discipline that traditionally has restricted itself to pure theory illustrates the broadly interdisciplinary implications of the major problems of our time, and the pressing need for scholars of conscience from all fields to apply their talents to solving them.

If even philosophers can, so to speak, dirty their hands in the soil, then surely geographers can do the same, taking care lest a growing tendency toward preoccupation with the artificial and the immaterial lead to excessive neglect of their traditional concern with the earth. By insisting on maintaining their special disciplinary advantage of working with their feet on the ground, geographers might even help to infuse into philosophy and general social theory, and perhaps even natural science, a much needed new sense of objective reality.

ACKNOWLEDGMENTS

The author wishes to thank Elbridge Rand, Steve Olson, Stan Dundon, William Friedland, Phil Wagner, and John Greene for helpful discussion and advice.

REFERENCES

Abbagnano, N. 1967. Positivism. In *The Encyclopedia of Philosophy,* 6:414–19. New York: Collier Macmillan.

Ackerman, E. 1963. Where is a research frontier? *Annals of the Association of American Geographers* 53:429–40.

Anderson, P. 1976. *Considerations on Western Marxism*. London: NLB.

———. 1984. *In the tracks of historical materialism*. Chicago: University of Chicago Press.

Arato, A., and Gebhardt, E., eds. 1987. *The essential Frankfurt School reader*. New York: Continuum.

Barbour, I., ed. 1973. *Western man and environmental ethics*. Reading, MA: Addison-Wesley.

Baumer, F. L., ed. 1964. *Main currents of western thought*. 2nd ed. New York: Knopf.

Bell, D. 1975. Lecture at Harvard University, March 1975.

Chappell, J. E., Jr. 1975. The ecological dimension: Russian and American views. *Annals of the Association of American Geographers* 65:144–62; also see commentary in *ibid.* 66:168–73.

———. 1979. Towards a logical electromagnetic theory. *Speculations in science and technology* 2:313–22, 338–40; also see letter in *ibid.* 3:488–95.

———. 1981. Environmental causation. In *Themes in geographic thought*, eds. M. Harvey and B. Holly, 163–84. London: Croom Helm.

———. 1988. Social Darwinism, pragmatism, and American geography. *History of Geography Journal* (formerly *History of Geography Newsletter*), 6. In press.

Chisholm, R. M., ed. 1960. *Realism and the background of phenomenology*. Glencoe, IL: The Free Press.

Dundon, S. 1986. The moral factor in innovative research. In *The agricultural scientific enterprise: A system in transition*, eds. L. Busch and W. Lacy, 39–51. Boulder, CO: Westview Press.

Duvall, B., and Sessions, G. 1985. *Deep ecology: Living as if nature mattered*. Salt Lake City, UT: Peregrine Smith Books.

Eiseley, L. 1958. *Darwin's century: Evolution and the men who discovered it*. Garden City, NY: Doubleday.

Friedland, W., and Kappel, T. 1979. *Production or perish*. Santa Cruz, CA: Project on Social Impact Assessment and Values.

Glacken, C. 1967. *Traces on the Rhodian shore: Nature and culture in western thought from ancient times to the end of the eighteenth century*. Berkeley and Los Angeles: University of California Press.

Gleick, J. 1987. *Chaos: Making a new science*. New York: Viking.

Gould, S. J. 1980. *The panda's thumb*. New York: D. Appleton.

Gregory, D. 1978. *Ideology, science and human geography*. New York: St. Martin's Press.

Griffin, D. R., ed. 1988. *The reenchantment of science: Postmodern proposals*. Albany, NY: State University of New York Press.

Harland, R. 1987. *Superstructuralism: The philosophy of structuralism and post-structuralism*. London and New York: Methuen.

Hartshorne, R. 1939. The nature of geography. *Annals of the Association of American Geographers* 29(3,4):entire issues.

Hofstadter, R. 1959 [1955]. *Social Darwinism in American thought*. Rev. ed. New York: George Braziller.

Hubbert, M. K. 1963. Are we retrogressing in science? *Geological Society of America Bulletin* 74:365–78.

Husserl, E. 1931. *Ideas*. Trans. W. R. Gibson from German ed. of 1913. London: George Allen & Unwin.

Idso, S. B. 1981. Climatic change: The role of atmospheric dust. *Geological Society of America Special Paper* 186:207–15.

———. 1985. The search for global CO_2 etc. "greenhouse effects." *Environmental Conservation* 12(1):29–35.

Ingold, T. 1986. *Evolution and social life*. Cambridge: Cambridge University Press.

Jackson, P., and Smith, S. J. 1984. *Exploring social geography*. London: George Allen & Unwin.

Kropotkin, P. 1972 [1914]. *Mutual aid: A factor of evolution*. 2nd Rev. ed. New York: N.Y.U. Press. (First ed. 1902.)

Kuhn, T. 1970. *The structure of scientific revolutions*. 2nd ed. Chicago: Univ. of Chicago Press. (First ed. 1962.)

Lappé, F. M., and Collins, J. 1977. *Food first*. Boston: Houghton Mifflin.

Leighly, J. 1955. What has happened to physical geography? *Annals of the Association of American Geographers* 45:309–18.

Lichtheim, G. 1961. *Marxism*. New York: Praeger.

Megill, A. 1985. *Prophets of extremity: Nietzsche, Heidegger, Foucault, Derrida*. Berkeley: University of California Press.

Relph, E. 1970. An inquiry into the relations between phenomenology and geography. *Canadian Geographer* 15:181–92.

Rostlund, E. 1956. Twentieth-century magic. *Landscape* 5:23–26.

Samuelsson, K. 1961. *Religion and economic action: A critique of Max Weber*. Trans. E. Geoffrey French from Swedish ed. of 1957. New York: Harper & Row.

Sauer, C. O. 1963 [1925]. The morphology of landscape. In *Land and Life,* ed. John Leighly, 315–50. Berkeley and Los Angeles: University of California Press.

Spencer, H. 1892 [1850]. *Social statics*. Abridged and rev. ed. New York: D. Appleton.

Spiegelberg, H. 1965. *The phenomenological movement*. 2 vols. 2nd ed. The Hague: Martinus Nijhoff.

Tawney, R. H. 1946 [1937]. *Religion and the rise of capitalism*. Rev. ed. New York: New American Library. (First ed. 1926).

Weber, M. 1930. *The Protestant ethic and the spirit of capitalism*. Trans. Talcott Parsons from German rev. ed. of 1920. New York: Charles Scribner's Sons. (First ed. 1904–5.)

———. 1947. *The theory of social and economic organization*. Trans. A. M. Henderson and Talcott Parsons. New York: Collier Macmillan (Free Press of Glencoe).

White, M., ed. 1955. *The age of analysis*. New York: New American Library.

Wild, J. 1948. *Introduction to realistic philosophy*. New York: Harper & Row.

Wilson, E. O. 1978. *On human nature*. Cambridge: Harvard University Press.

Woldenberg, M. 1986. Quantitative analysis of biological and fluvial networks. In *Microvascular networks*, ed. A. Popel and P. Johnson, 12–26. Basel: Karger.

Presenting with Pizazz: Oral Traditions in Geography

Larry Ford

Anyone who has attended a number of sessions at a large professional meeting has at some point been bored and disillusioned. Why, we say to ourselves, can't the speaker get to the point, use clearer language, and project legible overheads? The fact is that few of us have ever been trained or even been given much advice on how to present a paper effectively. Our training is focused on content rather than delivery, and on the written product rather than the oral event. This creates a problem for graduate students, since they usually present before they publish. Presenting that first paper at a professional meeting is a rite of passage for which preparation is crucial, but often lacking.

AXIOMS FOR PRESENTING

I shall divide these words of advice into two parts. The first consists of ten "consensus axioms" for presenting papers at professional meetings: rules that nearly everyone would agree with in principle if not in practice. The second part consists of a set of personal observations based on my own experiences as both a presenter and a listener. If the guidelines in both parts are followed, I can almost guarantee that people genuinely will want to listen to your paper and will not begin to read their programs, wondering whether they should bolt to the session across the hall instead.

And now for the axioms.

1. *An oral presentation is not the same thing as a written paper read aloud.* Many speakers simply take a part of their thesis or a paper they have written for publication and "present" it. The trick question, of course, is how do you "present"? To communicate information to a sleepy mob, some of whom may have been drinking heavily while others are trying to sort out the forty papers they have already heard that day, you should reorganize your material to emphasize a few main points in a clear and lively manner.

There is no consensus whether papers should be read verbatim or "talked through" from an outline. The very best papers I have heard were most often read by individuals who knew how to prepare and deliver a written text. Many of the very worst presentations have also been read—but by those who brought a tome not prepared specifically for oral delivery and then read from it looking down, as though no one else were in the room (after a time, no one was!). This is a matter of individual style, but, in either case, the paper and/or outline should be designed specifically for oral communication. If you can read clearly and enthusiastically while maintaining eye contact with the audience, you may find that this is the best way to say precisely what you want to say in exactly the time allotted. If you cannot, then speak from an outline.

2. *An oral presentation should be designed for the time allotted.* Most geographers who have presented a number of papers have had the experience of hearing the session chair whisper, "Pssst, pssst—two more minutes" or, "Please wrap it up." But it has constantly amazed me, when I have chaired sessions, that some speakers seem to be totally unaware of how long they have been speaking. On one occasion, I asked a presenter to please conclude. He replied that he was not even finished with the introduction—the paper itself had not yet begun! I now carry an alarm clock when I chair a session. It is not only rude to speak beyond the allotted time, but also it dilutes your message. Geographers have been conditioned to think in terms of fifteen- to twenty-minute packages at meetings, and they grow restless if a presentation lasts longer. Remember, the purpose of an oral presentation is not to present all your findings as they exist in your written product, but to use the time allotted to communicate enough of the information to get people interested in your work. That one final chart that you feel is essential may be the lullaby that puts the crowd to sleep. Time your material carefully in advance.
3. *An oral presentation should have a point.* It should have a clear purpose, tight direction, and a strong conclusion. Tired listeners want to know at the outset exactly what the paper is about and why they should want to know what you are going to tell them. It is generally not enough to announce that you are going to "look at some relationships" or "identify some problems" or "show some slides." Nothing makes an audience more restless than wondering what is the point of what they are hearing. Clear purpose and direction sometimes involve repetition: "the purpose of the paper is to show that," followed shortly thereafter by "these data show that," and so forth. A paper should have a strong conclusion as well, even if it is only tentative and not engraved in stone.
4. *An oral presentation should be tightly organized.* All the parts must be related, and the linkages should be emphasized, even if they seem obvious to you, the presenter. Nearly everyone who has listened to a large number of papers in one day, especially difficult or abstract papers, has experienced the "in and out syndrome," during which the listener follows the train of thought for a while and then mentally drops out to await the next clearly presented idea. What may be essential, elucidating backup material

in a published article, comes across as soporific obfuscation in an oral presentation, unless the linkages between thoughts are made explicit.

5. *An oral presentation in geography should be geographic.* While we wisely have become more accepting of ideas, theories, and methodologies that are clearly on the fringe of geographic pursuits, a common criticism heard at meetings is, "What does that have to do with geography?" This may be a real problem for students who are presenting findings from narrowly focused dissertation research, or based on a theory or methodology their adviser thinks is "on the cutting edge." The presenter must make explicit the relationship between the ideas or methodologies being discussed and geographic research. If this is not done, then only the already converted can participate in the ensuing discussion.

   This problem has become especially acute in recent years, with the rise of sophisticated technologies and "imported" social theories. It takes very little time and effort to demonstrate to the uninitiated why the topic is important to geography. If this is not done, there is the likelihood that the paper will contribute to the fragmentation of the discipline and that the speaker will have little impact beyond an established subgroup.

6. *An oral presentation should have a minimum of specialized jargon.* There are, obviously, a number of highly technical, specialized words and phrases that must be used in any sophisticated presentation. And it is not always wise or possible to define terms so that everyone can follow. Still, where possible, if you can say it in clear, simple English, then do so.

7. *An oral presentation should be rehearsed.* Geographers who teach have the luxury of presenting ideas to a captive audience before arriving at a professional meeting. Sometimes it is evident that changes clearly must be made if we are to spare our colleagues from having the same glazed eyes and uncomprehending looks as our students. If you are not teaching (or are not teaching an appropriate class), practice on your colleagues or anyone else who will listen. Words, phrases, ideas, and linkages that make perfect sense to you in a written format, when presented orally are sometimes misunderstood or not understood very easily. If you have no other options (listeners), practice your paper alone so you can become comfortable with cadence and phrasing. Occasionally, presenters read papers as though they are seeing them for the very first time, stumbling over words and showing slides upside down or transparencies out of order. Sometimes this is due to nerves, but often it is due to a lack of preparation.

   Make sure that your charts, graphs, tables, and other visual aids can be read from more than three feet away, whether you use slides or transparencies. Do not display more detail than viewers can assimilate easily.

   For those who would like to rehearse seriously, many universities now have facilities for videotaping presentations, which allows you to watch yourself in action and take steps to improve your performance. I hesitate to recommend this for fear that some novice speakers might put off forever giving an oral presentation after repeatedly witnessing their "first try." Nevertheless, it is an option and can be helpful.

8. *An oral presentation should be given in a room that the speaker has inspected beforehand.* Anyone scheduled to present a paper should inspect both the room and the facilities beforehand, and this holds true for the session chair as well. If possible, attend an earlier session in the room where you are scheduled to speak, and sit in the back or at the side to investigate the acoustics. People will not enjoy your paper if they cannot hear you; nor will they particularly enjoy you shouting at them. Find out where the light switches are, and check whether the equipment you need is in working order by testing it during the break just before your session. For example, sometimes a slide projector that works for one tray will not work for another. Although there may be workers in each room at some meetings to help remedy mishaps, it is always frustrating to watch people take valuable time to discover that they do not know how to adjust an overhead projector, cannot find the remote control for the slide projector, or cannot read with the lights turned down.
9. *An oral presentation should be given with energy and enthusiasm.* This does not mean that you should necessarily jump up and down like a game show host, but it does mean that if you show little or no interest in your own topic, it is highly unlikely that anyone else will either. At a large meeting you are competing for attention, and a little tasteful messianic fervor helps to sell both yourself and your ideas. If you are not genuinely excited about your topic, perhaps you should choose another.
10. *An oral presentation must have content.* Unlike articles submitted for publication, most oral presentations are not reviewed or are reviewed uncritically, so criticism comes only after the presentation itself. As competition for jobs and exposure has become intense, so has the pressure to perform as soon and as often as possible. I believe this is a good development. Do not hesitate to give a paper just as soon as you have something to say, even though you may fear that it is not good enough. But remember that an oral presentation demands the same high standards of scholarship and research as a written paper, even if some of the best parts must be left out in order to present the big picture. If you have nothing to say, don't say it.

Few geographers would argue with the above axioms. In a nutshell, the axioms simply say you should "be prepared": do your research, identify an appropriate amount of information for a twenty-minute talk, write a paper or outline specifically for an oral presentation, carefully prepare and organize any visual aids beforehand, clearly state your purpose and reemphasize it with a strong conclusion, check out the room in advance, and then "go at it" with enthusiasm.

There are, however, several other aspects of presenting a paper that are more a matter of personal style and preference. Nevertheless, I recommend that students planning to present a paper at least give some consideration to the following "world-according- to-Ford" guidelines. If they work, fine; if not, then they may help you develop a different kind of personal style. The important thing is not so much to follow my suggestions to the letter, but to spend some time and energy thinking instead about how to develop an effective presentation style. Sometimes this takes work, but it's worth the effort.

GUIDELINES FOR PRESENTERS

1. *Present a mixture of theory and empirical findings.* With few exceptions, the papers I enjoy most are those that utilize a strong (geographic) theoretical context and apply it to a clearly relevant case study. In geography, place is always a key variable. When someone discusses theory but fails to relate it to particular places, geographers find it very difficult to get the full picture and reach a deeper understanding of what the speaker has in mind. This is less important with written works because we can scan the references, ponder possible applications at our leisure, and do additional reading. In a "one-shot" oral presentation, however, we must grasp the theory quickly or not at all. Considering the many "imported" theories that find their way into geography meetings, it is important that presenters take the time to illustrate precisely how an idea can be applied to geographic issues in real places.

   Often the problem is not excessive theory, but lack of theory, or at least of explicitly presented theory. The speaker displays a plethora of maps showing the changing distribution of something, or shows slides illustrating the changing landscape of some place, without a clearly presented theoretical context. We need to know why things are happening, not merely that they are happening. A good paper quickly and concisely introduces a theoretical context, and then clearly illustrates how the theory contributes to our understanding of a real place (or a type of place). Imported theory is most relevant when the speaker can demonstrate that it has geographic applications. Similarly, place description can be trivial unless there is a point, an explanation of process in a theoretical context. Oral presentations should be hybrids: theory and case studies skillfully blended.
2. *Include a visual dimension in your oral presentation, whether or not the visual element is a main focus of the talk.* Professionals in many other fields, such as business and architecture, realize that graphic aids help, although sometimes they overdo it—picture a huge pad of paper on a tripod, with the word "first" scribbled on the first sheet. Nevertheless, well-designed illustrations and graphics can help us focus on even the most complex philosophical issues.

   The trick is to use visuals without overusing them. In the best presentations, the visuals stand alone as an added dimension of the talk, rather than as objects to be discussed and explained. It is not necessary to dwell on, or even mention, every slide if they obviously contribute to the ideas being discussed. Similarly, it is not necessary to show a graph or a table and then laboriously explain every trend-line or percentage. If these items are clearly presented, they should stand alone. The presenter can then simply say that "improvement occurred in every case," while the legible diagram on the screen shows the exact increases. Visuals enable a speaker to do more than one thing at a time, and thus are a valuable way to maximize the impact of a short presentation.

3. *Make your oral presentation appeal to the widest possible audience.* While there are many ways to achieve this, the easiest is through the use of maps. In a diverse field, with a wide variety of theories, philosophies, methodologies, and techniques, it is often difficult for geographers to communicate with one another. Especially, but not exclusively, in cases where it is impossible to provide everyone with a common basis for understanding the material being presented, maps can be invaluable devices for bringing people with diverse backgrounds together.

   Although I do not subscribe to the belief that "if you cannot map it, it is not geography," I do believe that if a topic can be mapped, it can be discussed intelligently by a large number of geographers. Maps allow us to contribute our own explanations and theories to the work being presented. I am often amazed at speakers who design presentations that provide a theoretical explanation for spatial patterns they never show us. We need to have a shared starting point, and in geography the map is usually the most appropriate beginning. If maps are neglected when they could have been used, valuable information is lost. If maps are inappropriate for a topic, then the presenter must come up with other ways to demonstrate that the topic is geographic.

4. *Discuss your methodology, but don't make that the entirety of your presentation.* In fact, the discussion of methodology should constitute only a very minor part of the total presentation. The most important parts of an oral presentation are the theoretical context and the results of the study. Methodology and data sources should be mentioned so the listener can understand in general how the study was organized and where the data came from. Normally, that is all that is needed, since very few listeners are interested in a detailed discussion of the strengths and weaknesses of a particular algorithm, or the difficulties in gaining admission to a library in Chad, or whatever—even though those issues currently may be of great importance to the presenter. Explain in a paragraph or two how your information was gathered and analyzed, and then move on to your findings and their significance. Methodology is best discussed after the session, with those who are particularly interested.

   You should, however, be prepared to discuss methodology. If questions arise in a discussion after the paper has been presented, or if people corner you in the hall demanding deeper insights, you should be able to defend your approach and your techniques. However, it is not wise to assume such interest exists. People normally have little interest in methodology unless they are already interested in the study and its findings. Give them the big picture first, and deal with the statistical quirks later.

Highly detailed methodological problems can often be discussed only in writing, so plan to communicate by mail with those who are doing similar research.

5. *Don't be afraid to arouse controversy.* Indeed, a good oral presentation should encourage at least a modest amount of controversy. One of the main reasons for presenting a paper at a professional meeting is to get feedback before you submit it for publication. If you present your results in an overly cautious manner, with only very tentative conclusions and rather vague problem statements, no one will be motivated to take issue with what you say. "Thank God," you may reply, but is that really what you want? As long as the debate is reasonably civil and amiable, differences of opinion can lead to better, more complete inquiries. If your ideas can be improved before your dissertation is finished and before you send off those first papers for the long and frustrating journal review process, why not present them clearly and forcefully?

   One problem with an overly cautious presentation is that it puts too heavy a burden on the audience. When speakers present a correlation and then proceed to say, "I don't know, it could mean this or it could mean that," it suggests that they have not thought about it sufficiently. After the data are collected and analyzed, you must then interpret them, even if this means introducing obvious subjectivity. Tell us what you think. Make a statement. Say something! We all grow by interacting with others.

   If you are truly concerned that making a strong conclusion will cause you to appear foolish, network beforehand. Send a copy of your paper (or even just your abstract and conclusions) to others who work in your field. Try to get some feedback and discussion before the event. Presenters who conclude by saying, "More work needs to be done on this topic," are almost guaranteed to elicit a chorus of yawns.

6. *Work to develop a personal, charismatic presentation style.* While many stylistic aspects of a good presentation—speaking loudly and clearly, making eye contact with the audience, being enthusiastic—were noted in the axioms given earlier, others are less obvious. For example, language should be not only concise and free of jargon, but interesting. Many of the great historical quotations are simply average thoughts phrased in an interesting and memorable way. I am not advocating flowery, poetic excess; rather, lively and interesting prose. We too often let our data speak for themselves and in the process rely heavily on trite and simply boring words and phrases. Buy a dictionary and a thesaurus. Read the works of writers like James Thurber and James Joyce. Give some careful thought to how you might use language to make an idea come alive.

   Many have suggested that humor is an important part of an oral presentation, but this is tricky business. A certain amount of tasteful wit is part and parcel of using interesting language, but we must be careful not to exceed our limits. If information can be communicated effectively by adding a touch of humor, or if a humorous remark can illustrate a relationship, all the better. However, we are not in the business of competing for laughs; we are not stand-up comics doing "a funny thing happened on the way to the convention." Today, we are probably erring on the side of too little rather than too much humor. Some speakers come across as God's scientific attaché on earth, and a little whimsy here and there might be good.

7. *Plan for the unexpected.* A good oral presentation should be flexible without sacrificing content. If the speaker ahead of you takes too much time, if the projector bulb burns out midway through your slides, if the audience is clearly squirming during your introduction, try to make immediate adjustments in your presentation. Sometimes it is useful to have a long version of your presentation—in case someone in your session fails to show up—and a short version—in case the session really gets behind schedule. This means having expendable sections of the paper, which can be expurgated without sacrificing the logic and flow of the talk. It means being able to adjust quickly and to give one example instead of two or three; it means arranging slides so that an earlier stopping point can serve as your conclusion, and you will not have to become flustered and rushed.

   If you are not the first speaker in your session, try to read the audience during the other presentations. You may decide that you have too many overheads or that a clearer problem statement is in order. Even if you are reading your paper verbatim, you can still make note of expendable paragraphs and/or possible changes in tempo and style.

8. *Keep your presentation fast paced.* By speaking as rapidly as you can without sacrificing clarity and organization, you accomplish several things. First, you can communicate far more in the time allotted, simply by uttering more words per minute (how's that for a scientific finding?). Second, a rapid delivery contributes to a sense of excitement and enthusiasm and can give the topic a sense of urgency. Third, by combining a rapid pace with interesting phrasing, good visuals, regular eye contact, and a minimum of pauses and gaps, you can prevent the audience from mentally "dropping out." They simply will not have time to start thinking about anything except what you are saying. A fast pace also facilitates flexibility, since it allows you to move through topics at different speeds to get to where you want to be.

9. *Actively encourage audience interaction through verbal and nonverbal communication.* Be confident but not arrogant; be willing to defend but not defensive; be open to suggestions but not too willing to follow trendy ideas. Indicate in every possible way that you sincerely would like to discuss your ideas and would appreciate constructive interaction. Some geographers present finely crafted papers at meetings, but they do so with a posture and stance that lead the listener to believe that the issue is settled and that the speaker wants no advice. If this becomes the norm, why have meetings at all?

10. *Attend meetings and listen carefully.* Find out who your colleagues think are good speakers, and attend their sessions regardless of their topics. Take notes on elements of the presentation, such as use of visuals, pace, audience attention, organization, and phrasing. Find out which speakers you admire, and learn from them. Conversely, find out what annoys you,

and learn from that. I have heard people complain vociferously about incompetent and boring speakers, only to watch these same people perform in a similar fashion shortly thereafter. There is always room to learn from one another.

Unfortunately, content will not stand alone. Research, no matter how well done, and insights, no matter how scholarly, need to be presented effectively to have maximum impact. For those planning to present a paper, the real work begins when the research is completed.

Finally, there are a few bits of advice that are even more personal and subjective than those already offered. How to dress, for example, is an important issue for some presenters. My advice is to dress in a manner that enables you to feel confident and good about yourself. In general, I would advise dressing attractively, appropriately, and comfortably. Appropriateness is often a debatable issue: I am not the dark-suit type myself, and I once gave a paper in Hawaii wearing a large orchid in my hair. Nevertheless, dressing neatly can enhance your presentation.

There are other guidelines that are quite optional, at least at the livelier conventions. For example, some argue that it is advisable to get more than four hours' sleep the night before a presentation and to avoid hangovers. I am still weighing this advice.

PREPARING AN ABSTRACT

The first thing you have to do when planning to present a paper at a meeting is to prepare a title and abstract. For some this represents a real challenge, especially since abstracts are usually required to be very short and yet say so much. There are two very different kinds of abstracts. First are those that summarize work already completed. These are relatively easy to write. Second are those that must be written before the results are in and therefore must include a bit of deliberate vagueness here and there. These are harder to write. Still, they must be written.

A perusal of the Association of American Geographers annual meeting abstracts shows that—surprisingly, I think— there is no consensus at all as to what an abstract should include. Some give long, pleading justifications for the study, as though the author were seeking permission to do it, but virtually nothing in the way of results. Others emphasize methodology and/or theoretical framework in order to clarify the type of "lens" through which the data will be viewed (once gathered). Still others jump right into the results, beginning with, "I found that. . ."

I believe that the best abstracts are well balanced, providing some information from all these categories. A good abstract should begin with the theoretical context or conventional wisdom being examined in the study. It should begin with a phrase like, "According to such and such theory," or "Much of the work in such and such field assumes that. . ." Next, the abstract should include a problem statement demonstrating how and why the researcher plans to test the theory or prevailing wisdom. A sentence or two on methodology and data sources should follow. The main body of the abstract should give the "big picture" results of the study. A final statement should show how the findings (or expected findings) relate to the context. Obviously not all abstracts can follow this model, since some papers deal exclusively with methodology or philosophical issues, but it does provide a possible norm. Each section of the abstract can contain only a sentence or two, so above all the abstract must be concise. Given the immense number of abstracts in many program guides these days, the shorter ones may be the most likely to be read.

SO, WHY GO THROUGH ALL THIS?

Why present a paper at a meeting? First, there are some very good nondebatable reasons, such as your adviser telling you to do so. But beyond that there are some very real and worthwhile rewards. Presenting your work allows you not only to get immediate feedback in the form of additional sources of information, methodological innovations, and deeper insights, but also to meet new people and make new friends. Learning to "network" is an important part of both the academic world and the diverse world of applied geography. Without an introduction, it is often difficult to meet people quickly at a huge meeting. Having someone come up to you and say, "I really enjoyed your paper," and going on from there is really what professional meetings are all about.

Entering Academe: The Search for Jobs

William Wyckoff

There are many good reasons to avoid an academic career in geography. Even after you land that first job, the pay is often low, and the hours are always long. The relative aging of the North American population and the predicted declines in future student enrollments are other long-term imponderables. There is the capricious spirit of the field itself: geography has a tendency to shift its emphasis. Love affairs with new ideas and approaches can be surprisingly short and unpredictable; today's hotshot in geographic information systems (GIS) may become tomorrow's high-tech has-been. However, the hallowed halls of academe still beckon many of us into long and satisfying careers in university teaching and research. Despite the caveats, we wouldn't have it any other way.

Clearly, however, those initially attracted to an academic career should ponder the decision carefully. It is not for everybody. Like many professions, academia demands a great range of skills that are never written down or codified. Therefore, graduate students contemplating a university career should talk it over at length with faculty, especially their advisers.

Participate actively in your department's activities. Help give and organize seminars, entertain guests, and contribute to the department's *esprit de corps*. Initial exposure to departmental life may give you clues as to whether or not to pursue an academic job. Carefully consider your other employment and career prospects. Geography outside the university will almost certainly grow more rapidly than geography within.

Still determined to set your course for academe? A few general comments on long-term employment strategies you can pursue as a graduate student are appropriate and may significantly increase your chances of quickly landing your first academic position. There are several things you can do to increase your professional visibility before you ever apply for that first academic position. Work with your adviser and other faculty members; plan poster and paper presentations for regional or national conferences of the Association of American Geographers (AAG) and the Canadian Association of Geographers (CAG) or for more specialized meetings in your own subfield. Don't present research just for the sake of presentation, however. Select only your *best* work. It may represent thesis or dissertation research or a strong seminar paper that a faculty member encourages you to develop further. After all, potential employers, even a year or two down the line, may be in your audience. Develop your professional persona early. Scholarly meetings are an excellent place to show your skills, and your participation will be viewed later as a sign of initiative and potential.

Publishing your research is another excellent indication of promise. You may consider publishing work jointly with your adviser or other faculty. Short pieces might be appropriate submissions to the *Professional Geographer*, the *Journal of Geography*, and other regional publications. Longer work published in the *Geographical Review*, the *Annals*, the *Canadian Geographer*, or other specialty publications will be an even bigger feather in your cap. Keep in mind, however, that acceptance of articles by most scholarly publications is dependent upon a rigorous peer review process, so don't be disappointed if your initial efforts are unsuccessful. Even evidence of submitted (if not yet accepted) work may help convince a potential employer of your enthusiasm and demonstrate a positive, aggressive attitude. Grant applications—dissertation fellowships from the National Science Foundation, special departmental or university funding, and so forth—suggest a similar keenness, and your hunger for funding would hardly disturb potential "bosses," such as department heads and deans.

Some final words before we plunge into the nasty Darwinian details of the job search. Do not be surprised at the length and frustration of the process. Many new, young professionals in a variety of fields can expect an active job search to take six months to a year. The process can be significantly longer than that in academe, especially to land that most coveted of slots, a tenure-track position. So, be patient. Cultivate a positive attitude, and don't let the natural stress that accompanies the process overwhelm you. Be assured of this: if you are very stubborn, very good at what you do, and develop a very solid record of promise as a graduate student, your chances are excellent for opening the door to academe. Obviously, it takes much planning, a lot of hard work, and maybe just a bit of luck. But, before you can open that door, you need to know how to knock.

APPLYING FOR AN ACADEMIC JOB

Finding the right job is the first step to getting the right job. The search for available positions should involve a thorough investigation of several resources. Best known to North American geographers are positions advertised in the *AAG Newsletter*. That is a reasonable place to start, but do not ignore the *Chronicle of Higher Education*. Some universities, especially smaller schools, may choose to advertise only in the latter. Also, make use of your department and its contacts. Many graduate departments maintain a job file that will contain academic positions. These notices often are mailed several weeks before they appear in the *Newsletter* or the *Chronicle*. Cultivate direct contacts with your adviser and other faculty members. Let them know you are interested in finding an academic position. They will provide valuable help in gathering information about possible openings. Finally, participate in national AAG and CAG conferences, and make use of the AAG Convention Placement Service. Jobs that open suddenly in the spring are sometimes advertised there and nowhere else.

You have several important decisions to make as you initiate your job search. First, there is the issue of self-description. What kind of geographer are you? What are your special assets and skills? A clear conception of what you are best trained to do will lead you more quickly to the jobs really meant for you.

Second, you must decide whether to consider temporary, one- or two-year appointments in addition to tenure-track positions in your search. Obviously, most job candidates prefer the latter, but temporary jobs have at least three decided benefits: (1) they are often easier to get at the entry level; (2) they provide excellent experience to build your teaching record; and (3) they may lead directly to tenure-track appointments in the same department. Pitfalls also exist, however, and you should not ignore them: the pay is often low; teaching loads are typically heavy, making new research and writing difficult if not impossible; and relocation costs can be considerable. A long string of one-year appointments is not the way to professional success. Short-term professional survival may prove to be your long-term professional demise.

Third, you may need to weigh other geographical and economic variables that may affect and limit your job search—e.g., you want to live in a particular region, or your spouse requires access to a particular industry or employment situation. You should ponder all these variables as you begin to scan the listings of jobs available.

As you peruse job descriptions, read each word carefully. Every key descriptive term should give you some insight into what may be a department's very complicated search process. Rarely is there complete agreement over a job description. Haggling and compromise typically produce the final notice, and your careful reading of that description may suggest an important but unwritten "subtext" to consider as you apply. Also, it is certainly proper to contact the search committee with general questions concerning any job description.

Ask yourself whether the description really fits you. Do not expect to find an opening that is so perfectly tailored that it advertises for no one but you. You must be flexible. On the other hand, much wasted effort goes into applying for jobs for which applicants are not really qualified. If you are a coastal geomorphologist, it makes little sense to apply for a job clearly earmarked for an economic geographer, even if you have had a class or two in location theory. Your record will largely speak for you, and every job description will typically generate a considerable number of qualified applicants. Concentrate your efforts where they will be likely to produce results.

Assembling the required materials for the job application is the next step. Most potential employers will ask for several basic pieces of information, including a letter of application, a curriculum vitae, and several (usually three) letters of recommendation; they may also request transcripts, especially those reflecting your graduate school experience.

Your letter of application should accomplish several tasks. Most importantly, it is the sole direct and personal link between you and the job described. Your letter should make that connection explicit by discussing your own qualifications in light of the job description. Tailor every letter to a specific job. Keep the letter fairly short, usually one to one and one-half, single-spaced typed pages. Don't bore your reader with your life story. Instead, demonstrate the specific links between your abilities and the department's needs. Discuss your teaching experience and philosophy. First-time job applicants buried in doctoral research often forget that departments really do need dynamic and enthusiastic teachers as well as researchers and are looking for evidence of teaching potential. Briefly describe your research interests and accomplishments. Also, indicate your broader and longer-term research agenda and, if possible, suggest ways your research background may complement the orientation of the department to which you are applying, the university, or even the region. Do not forget to include the names of those writing your letters of reference. Finally, have the letter proofread for errors after you have typed it neatly on quality paper. Wave your hand over the sealed envelope a few times for good luck, and drop it in the mail along with your curriculum vitae and any other required materials.

The curriculum vitae (usually called the c.v.) is another critical component of your job application. This venerable Latin term literally means "the course of one's life." The purpose of the c.v. is to provide a short account of your professional career and to outline your general qualifications for academic employment. Precise layouts vary, and you should feel free to consult with your adviser and other faculty members to select a method of organization that is visually attractive and that best highlights your own strengths. The first thing that may strike you is the longer length of the typical c.v., compared to a nonacademic résumé. This longer format is acceptable because it permits a detailed listing of professional activities over many years. This can add up to a dozen or more pages for senior scholars. Don't be discouraged if your c.v. is considerably shorter. Employers looking at entry-level applicants are not going to expect page after page of accomplishments. What they do expect to see, however, is promise and professionalism. They will also likely be looking for someone with a well-rounded geographical background who, though inexperienced, already demonstrates initiative and a willingness to become involved professionally.

Every c.v. should contain certain basic pieces of information. Your educational background should be given on the first page and should include material on degrees granted and/or pending. You may also want to provide brief titles of your thesis and dissertation work. Make special note of any academic awards and scholarships. If you can cite several, devote a special section of your c.v. to academic awards.

In another category, you should describe your teaching background. Include all positions in which you have served as an instructor, and briefly list the titles of the courses. Also include teaching assistant assignments and even tutoring jobs if your teaching experience is limited. To fortify your teaching background while still a graduate student, explore the opportunities for teaching in your department and in other nearby universities and schools. Ideally, you want to demonstrate an already proven ability and desire to get in front of the classroom.

In another section of the c.v., detail your research background and record your professional activities. This may include a variety of accomplishments. Highlight any publications, especially articles in refereed journals. They are important signs of professional development and legitimacy. Also include short reviews and commentaries that you may have published. You may want to note the titles of additional papers submitted and provide a brief description of ongoing research projects. List professional papers and posters you have presented at scholarly meetings. Describe any grants you have applied for, and note any successes. Each of these items will suggest your dedication to an active and continuing research career.

Additional professional experience and background can be detailed in an accompanying section. You may want to note service activities in regional or national associations. Examples of additional professional experience include membership in scholarly organizations, organization and participation in conferences or activities, and involvement in special seminars and workshops.

A few final words on preparing a c.v: Be honest. Do not claim to have accomplished what you have yet to do. Ponder the format carefully and choose one that best fits *your* needs. Work to create a visually pleasing document that is balanced between headings and margins. Leave blank space on the page to soothe the reader's eye. Remember, you are not constrained by length, as is the case with most nonacademic résumés. Print the c.v. on quality paper, and choose an attractive and readable typeface. Have someone else proofread your final document for typographical errors and, above all, don't hesitate to show your first attempts to other colleagues and faculty *before* you send them to prospective employers.

Solicit your letters of recommendation in a professional manner. Ask for them in writing from your adviser and other faculty members. Provide copies of the job description and your c.v. to help your references in making the case that you are not only qualified for a prospective job—so are many others—but that you are the *best* qualified. Be selective in choosing individuals to write your letters. Try to make good matches that may reflect personal and professional links between your references and your potential future department. And don't overuse and exhaust your advocates; their enthusiasm for you may begin to wane if you burden them with too many writing chores. It is also wise to send precisely the required number of letters—no more, no less. Don't blitz the search committee with six letters of recommendation when it has asked for only three.

Supplemental materials may include reprints of your publications or abstracts if they are requested. If they are not requested, don't overload the search committee with extra bulk in your application. An exception might be an extremely attractive and polished piece of published work that is available as a reprint and that can easily be slipped in with your application materials. You might also want to include a brief summary of your graduate coursework if it provides additional evidence of your special skills, background, or experience.

Now comes the hard part. You must wait for the search committee to make its initial series of decisions regarding possible interviews. The wait can easily last two months beyond the advertised closing date of the position. You will probably receive a brief letter acknowledging receipt of your application, and then silence. A complex internal review process is now taking place as the search committee assembles files, makes initial cuts, and presents potential candidates to the faculty. A list of seven to ten candidates is sometimes produced, with polite and short rejection letters sent to the remaining applicants. Those still under consideration are sometimes informed of their status, sometimes not. Eventually, the search committee decides on a *short list*, often three candidates, to bring in for on-site interviews. If the hiring process is coincident with the AAG annual spring meeting, an initial screening, often conducted in small, overcrowded hotel rooms, takes place at the convention. You can do little during this entire period. You may call the search committee and ask for a general update on your status. You may also provide them with significant and relevant additions to your c.v.; for example, you might let them know if you have completed your degree, published a paper, or received a grant. This information updates your file and keeps your name in front of the committee.

INTERVIEWING FOR AN ACADEMIC JOB

The telephone rings. Suddenly you are in the midst of a conversation with a potential employer. You have been selected for an on-site interview. This initial conversation, usually with the department head, should establish possible dates of travel and include a general discussion of planned activities for your visit. Do not be afraid to ask questions and to clarify how travel is to be arranged and paid for. Try to be flexible with your own time schedule to accommodate the department's schedule. Often, the search committee has a limited number of open time slots for interviews and/or is under pressure to reach a decision. This leaves the committee with little flexibility.

Before you board the plane, you need to accomplish four important tasks:

1. You must learn something about your potential department. There is nothing devious about this; it is simply a solid professional strategy that will pay handsome rewards upon your visit. Know something about the department's program and the research interests of the faculty, especially those in your subfields. Think about potential parallels and building blocks between your own background and those of the faculty. What can you add to their program? How will you fit in with existing personnel?
2. Learn something about the region where the job is located. After all, you are a geographer, and your potential employer will appreciate the fact that you exhibit a sense of excitement and interest in your new setting. Be enthusiastic about how your move will serve as a professional opportunity to expand your research interests in a new region or with a new faculty and university. Potential employers are looking for individuals who will "fit in," enjoy their new home, and contribute actively in their new regional and institutional settings.
3. Most important, focus your efforts on assembling a coherent, exciting, and impeccably prepared talk for your upcoming interview. Typically, you will be asked to present a forty- to forty-five-minute seminar or colloquium to faculty and students. You may also be asked to give a lecture to a class. Clarify ahead of time the length and the audience of these presentations. For a research seminar or colloquium presentation, choose a topic that you know well, one that can display your talents as a teacher, and one that allows you to present your research in a clear, crisp, exciting, and intellectually stimulating fashion—all of which are easier to write about than do!

   Choose a topic from your principal research area, usually drawn from your thesis or dissertation. Time the talk to end about two minutes early.

Practice the presentation both alone and in front of others, until you feel confident and completely in control of the material. Tell your audience precisely what you are going to do, do it, and then tell them you have done it. If preparation time permits, use quality visual aids. Both slides and overhead transparencies are acceptable. Keep in mind that your audience is looking for three things: they want to know (a) that you can be *enthusiastic* (a good teacher); (b) that you can be *analytical and creative* (a good researcher); and (c) that you can *think on your feet* and are *intelligent* (a good colleague). Almost invariably, the presentation is a key component of the interview. Practice, practice, practice!

4. Try to handle the natural anxiety that comes with occupying center stage amid ten or twenty faculty members for two or more days. Some people deal with these fears better than others. Being *well prepared* and *well rested*—though these may be mutually exclusive!—are the two keys for a successful interview. Arrange mock interviews, and practice answering possible questions if you still are anxious. Your adviser and your university's career services office can give you additional suggestions.

You will have time to ponder some general interviewing tips and skills as you travel. Remember that multi-day interviews are often a test of endurance. Pace yourself throughout. Take advantage of breaks and quiet time if and when offered. During any interview it is important to "be yourself" and to appear at ease, but *not* laid-back. Although the atmosphere may indeed be relaxed and friendly, be at your *professional* best from the time you leave the plane till you catch your flight home. Naturally, your future colleagues will want to know that you are easy to talk to, but they should also want to know that you are a well-rounded and competent geographer.

Dress professionally. Jeans, T-shirts, and tennis shoes are fine for an afternoon jog and may even be the preferred attire of some of your future colleagues, but more formal dress is appropriate. It is more acceptable to overdress a bit than to underdress for an interview. You need not be stodgy or overly conservative, but a banana-colored suit or a slinky dress may leave your interviewers nonplussed. Use your common sense to select attire that is comfortable, reflects your personality, and is appropriate for the occasion.

Be positive, aggressive, enthusiastic, and realistic with your potential employers. Watch for opportunities to point out where you might fit into the department. Show enthusiasm for the position, but do not agree to teach courses or take on assignments that seem unreasonable. You must be flexible, but your employers must understand what you are and are not prepared to do if you are hired. Often you can simply leave the door open to further discussion. It is possible that improper questions may be posed, such as "What church do you attend?" or "How does your husband feel about doing housework when you're at the office?" In such a case, be prepared to answer politely and professionally. It is probably unwise to make a scene over such inappropriate but often unintentional *faux pas*.

As soon as you arrive, the ritual of the academic interview begins. Remember that both you and your potential employer are about to play out a complex and time-honored series of well-orchestrated activities: you will have an opportunity to display your excellence, and your potential employer will have an opportunity to scrutinize you in a variety of professional and social settings. Over the next day or two, your interviewers will be looking—in faculty meetings, in the classroom, and at meals—for someone well qualified for the position who is also a competent geographer and a potential long-term colleague. Most academic interviews consist of some combination of the following eight activities. Be prepared for each.

1. *The Entrance Interview*. This is usually an opportunity for you to meet and get acquainted with the department head. He or she will probably fill you in on the details of your schedule, provide you with more background on the department, and pose some initial questions about your needs and wants as a potential employee. Depending on scheduling, the department head may pick you up at the airport or meet with you over breakfast to discuss such matters. Use this opportunity to ask any general questions about the department, position, or university.
2. *Meetings with Individual Faculty*. During your stay you will have an opportunity to speak with most of the faculty individually and/or in small groups. In a large department, this can be a tiring process. Here is where your homework and enthusiasm pay off. No doubt, you may answer the same questions about your research and teaching interests many times. Still, you must continue to generate excitement over your own background and prospects. Balance the conversation between a discussion of your interests and theirs. You can find out a good deal about life in a department by listening carefully to members' discussions of teaching, research, and service responsibilities. At the very least, try to keep all the new faces and names straight!
3. *The Tour*. Your visit should include a tour of the department and campus. Use this time to relax and take in the surroundings. This is usually an opportunity for your hosts to show off a bit and to describe the amenities of their professional setting. This may be a good time for you to ask about the quality of labs, the library, cartographic services, and computer facilities. If time permits, the tour may extend into the community as well. This will give you a chance to view housing, shopping districts, schools, and other local services. It is an excellent idea to take advantage of this opportunity or to do it on your own time during your visit. Your spouse will surely be interested in the community as well as the position in the department. In some cases, especially in dual-career families, your spouse may want to accompany you on an interview. Such arrangements, usually at your own additional cost, can generally be worked out with your hosts.
4. *Meeting with the Dean*. Depending on the size of the institution and scheduling limitations, you may have an opportunity to meet the dean or assistant dean of the college. Such encounters are usually pleasant, brief, and informational. Do not expect to be grilled. The dean will extend a welcome and discuss some of the advantages of working at the institution. He or she may also ask about your research interests, always mindful of potential grant-getting opportunities for the college as well as ways that your background might complement the university's broader research goals and agenda.
5. *Meeting with Students*. Often, you will have the chance to spend some time with undergraduate and graduate students in the department. This is an important opportunity for several reasons. Obviously, you will be able to

obtain a different perspective on the department and its program. Such a meeting also allows students to get to know you and to explore how you might fit their needs in the program. Remember, graduate students in particular may carry some weight in the decision making for the position. Be friendly and open, but be at your professional best. Meeting with students is simply another part of the interview process.

6. *The Presentation*. Simply put, this is often the key to the entire interview, especially when two or three candidates seem otherwise closely matched. Use it to your advantage by making it the high point of your visit. Come well prepared, and convince your hosts that your research and teaching potential are unbeatable. Take advantage of any quiet time before your presentation. If none is offered, request thirty minutes to catch your breath, make sure your slides aren't upside down, and relax a bit. Then, focus your energies on making an enthusiastic and thoughtful presentation, and answer questions and comments afterwards as completely and honestly as possible. Timing for presentations during your visit will vary. Although you may be fresher in the morning, busy schedules often mean the department must wait until mid- or late afternoon to meet. Save some mental and physical energy for this key event!
7. *Evening Social Festivities*. After a busy day visiting with faculty, you will probably spend a busy evening visiting with faculty. Typically, arrangements for dinner and/or a social hour are made to provide more informal settings for discussion. But beware—this is often the occasion where your hosts determine whether or not you "fit" as a potential colleague. You are being looked over carefully. If you do drink, let this evening be a demonstration of your moderation. Few inebriated job candidates are ultimately successful. Save the celebrating for after you receive the job offer. Use this evening to relax, to find out more about the university and community, and to assure your hosts that you are at ease in professional social situations.
8. *The Departure Interview*. Before you complete your visit, the department head will usually have a closing session with you to ask and answer any final questions. You may get some clues as to how well the interview has gone. The department head may discuss issues such as salary and other considerations that would help in making a decision should the position be offered to you. Unless you receive a firm offer at this time—and this can happen—be positive and upbeat, but do not share all of your thoughts about what you think is a fair salary and what would convince you to accept the position. Leave that to subsequent negotiations, if an offer is forthcoming. If indeed you are departing with a real desire for an offer, express your enthusiasm about the potential for working with your new colleagues, and assure the head that you would be quite interested in considering the position. If your visit has convinced you that the department is not for you, thank your hosts and depart. You can always turn down an offer if one is extended to you.

Following your on-site visit, send brief thank-you letters to the department head and to others who were particularly helpful. These visits take time and money, and your appreciation for their efforts, even if no offer is forthcoming, is simply professional courtesy. A second wait now ensues, and its length depends on how near the department is to the end of the job selection process. It is appropriate during the departure interview to ask about the general timing of the hiring decision. If that time passes, you should feel free to check with the department, especially if you have other interviews or offers on the horizon. Remember that a complex, idiosyncratic, and often highly political process that is totally out of your control is now taking place to select a final candidate. The faculty can deadlock. The decision can be delayed. One or two top candidates can decline the position, leaving the way open for an offer to you. Most department heads will inform you through a telephone call immediately upon the final acceptance of the position by another candidate.

Perhaps, however, the telephone call brings more positive news: You are extended an offer. Negotiations now begin to "close the deal." A few initial tips: Assume good faith on the part of the department head; his or her job now is to lobby for your needs. Be honest and reasonable about salary and other demands. Be patient, and work through the process professionally from beginning to end.

Obviously salary will be a major issue. Usually the department head has some flexibility, but you should not abuse it. The figure you agree upon will probably be somewhat less than you would consider ideal, and it will probably be somewhat more than the department head initially offers. It is proper and expected, however, to negotiate for salary, especially if you are above an entry-level position. Other issues may come into play, as well. It is acceptable to ask about payment of moving costs and to inquire about any "start up" costs and equipment (lab machinery, computer facilities, and so forth) you believe would be necessary at the outset. Often, the department head has more flexibility in handling these issues than in dealing with salary.

If you have teaching experience, you also need to discuss whether it is desirable or permissible to count your years of experience toward promotion and tenure in your new position. There are various schools of thought on this issue. If you are well along in producing publishable research, and if there are sound financial incentives to quick promotion, you may want to consider bringing one or two years of teaching experience with you. If you are only near the beginning of your publishing career, even if you have taught a year or two, it is often wise not to bring in years of experience. Ask your future boss, who probably has the best perspective: Your new department head knows something about you and about what the department and university demand for promotion and tenure.

It is proper, in most cases, to take up to a week or more to make your final decision once an offer has been extended. Remember, however, that departments are usually under some pressure to hire. Don't create delays unless extenuating circumstances (other offers) make it absolutely necessary. Don't be coy. You earn no professional respect by being less than forthcoming at this critical stage. Be honest about competing offers or about other circumstances, perhaps personal, which may cause you to hesitate.

At its best, any academic job search is a lot of hard work, often exhausting, and potentially discouraging. At its worst, you may ponder selling insurance, driving a cab, or playing professional sports as more appealing and likely alternatives. Remember that the competition is tough and that much of the process is completely beyond your control. Many excellent young professionals are not hired immediately. Ultimately, success is equated to some extent with sheer stubbornness. Do not be easily discouraged. While you await the ideal job announcement, there are plenty of professional (though often uncompensated!) activities you can engage in to augment your research and to demonstrate your desire to be a productive and enthusiastic geographer. Good luck!

FURTHER READINGS

Bolles, R. N. 1988. *The 1988 what color is your parachute?* Berkeley, CA: Ten Speed Press.

Fraser, J. K. 1983. Road less traveled: Reflections on a career in geography. *Canadian Geographer* 27:305–12.

Green, D. B. 1983. Teaching positions in geography in the United States: What specialties have been in demand? *AAG Newsletter* 18, 4:14–15.

Hausladen, G., and Wyckoff, W. 1985. Our discipline's demographic futures: Retirements, vacancies, and appointment priorities. *Professional Geographer* 37:339–43.

Mikesell, M. W. 1980. The sixteen-million-hour question. *Professional Geographer* 32:263–68.

Salter, C. L. 1983. What can I do with geography? *Professional Geographer* 35:266–73.

Scully, M. G. 1982. Academic geography: Few students, closed departments, fuzzy image. *Chronicle of Higher Education*. May 26, 1982: 1, 12.

BEYOND GRADUATE SCHOOL

Ascent From the Maelström

Robert A. Rundstrom

You have arrived. You are among North America's best minds. As a young professor, you are placed on a pedestal, elevated to the status of soothsayer, and perceived to have a rewarding and relatively easy life with plenty of time off for thinking great thoughts. The salary, by no means extravagant, nevertheless should keep you in scotch, bourbon, pipe tobacco, and tweed jackets for many years, as you spend your days quietly amid stacks of thoughtful reading material, free from the distractions of the mundane world.

> This was most unusual—something that had never happened to us before—and I began to feel a little uneasy, without exactly knowing why. We put the boat on the wind, but could make no headway at all for the eddies, and I was upon the point of proposing to return to the anchorage.[1]

As a graduate student inside the system, you have an inkling that the public perception of academia and its inhabitants, as described above, may be slightly askew, but there is precious little that prepares you for the vortex of centripetal forces you will encounter in your first two years working in a North American university. As a graduate student, you lead a relatively simple life with few responsibilities. Your mind is keen and inquisitive. You are probably at a top-notch university where geography is prominent and where you enjoy camaraderie and commiseration with your peers. To be sure, there are ups and downs; graduate school can be an emotional rollercoaster. And the federal deficit pales in comparison to your financial situation. Graduate school requires an oath of poverty, the ability to incur a "gigadebt" without a nervous breakdown, and a willingness to postpone much-needed medical and dental work for a decade or so. Nevertheless, these are quite likely your halcyon years. Hard to believe, isn't it? You think time is on your side. You don't realize that your first few semesters in a university faculty position—even if you are lucky enough to land a tenure-track appointment—will

[1]Edgar Allan Poe, "A Descent into the Maelström." *Selected Prose, Poetry, and Eureka,* W. H. Auden, ed. (New York: Holt, Rinehart and Winston, Inc., 1950), 139. Page citations for subsequent references will be noted parenthetically in the text. The title is a variation on Edgar Allan Poe's "A Descent into the Maelström."

be regarded in your old age as one of the hardest periods of your life. Having entered as an Acting Assistant Professor of Geographic and Cartographic Sciences—a tag nicely displaying the triple themes of false dignity, arrogance, and preposterous hyperbole that are too commonly encountered on college campuses—I know from experience that the early days as an assistant professor can be a dark void into which many academics fall. Luckily, most of them return, but little news of the experience reaches the ears of subsequent generations.

There are no support groups, free clinics, halfway houses, or "rehab" or "detox" centers for struggling young professors. But you do not need them. Remember, you are among the honored, the chosen. Your motives and mind are pure and strong. To be honest, however, there are some academic geographers who—even if they do not have a substance abuse problem or a relationship on the rocks—still have some obvious personal perversity such as wearing polyester, an obsession with fractals, an urge to sleep in congealed Jell-O, or some other idiosyncrasy that indicates they too, at one time, got sucked into the whirlpool. How did successful and respected geographers get so disturbed? What happens in the whirl of that first year or two? What can the young, aspiring academic honestly expect?

In this chapter I will attempt to provide a sense of this warping experience. Many kind souls will tell you the lighter side, casting it in terms that bear no relation to reality as experienced. But the true depth and darkness of the void need direct illumination. I make no pretense of knowing the correct course. But I *am* saying this is the way it *is*. Like all good scholarship, of course, my remarks are perfectly objective, without personal bias, reflecting no attempt at self-aggrandizement, made with reverence at all the appropriate junctures, and bereft of any emotional involvement that might twist their meaning—or nearly so.

THE DESCENT

> . . .the terrific funnel, whose interior, as far as the eye could fathom it, was a smooth, shining, and jet-black wall of water, inclined to the horizon at an angle of some forty-five degrees, speeding dizzily round and round with a swaying and sweltering motion, and sending forth to the winds an appalling voice, half shriek, half roar, such as not even the mighty cataract of Niagara ever lifts up in its agony to Heaven (p. 134).

The first few semesters in a faculty position are not even remotely related to what many call "the transition period." The temporal and spatial gradient traversed is precipitous, a steep descent into uncharted, unrecognizable, and turbulent waters. The speed of the descent and the dense mist in the funnel make navigation difficult, which is perhaps the reason precious few veterans can offer insights or advice on its transit.

Your first university appointment signifies crossing a career threshold, and it is endowed with all the qualities typically associated with the passage from one social group to the margin of another. There is a sense of having little control over the new forces shaping your life and of not being able to characterize these forces precisely. It is an old phrase, but a true one, that little in graduate school prepares you for academic life, and what you did learn doesn't apply. It is not surprising, then, that most faculty newcomers are left to their own devices, merely to get through in the hope that something better waits on the other side or, more accurately, at the surface.

You will form new concepts of money, students, colleagues, administrators, teaching, curricula, research, and time. These form the varied but powerful streams of the maelström. The vortex tightens if your job is in a weak or small department or in a nongeography department, and it becomes tornadic if you take a position without having finished your dissertation. Today's market seldom gives you the luxury to choose the institutional setting you desire, and getting bread on the table is important. In short, you "do what you gotta do." You take the job and feel the tug of the current.

Tossing Pennies in the Goldfish Pond

One of the great attractions in obtaining a teaching post in a modern, North American geography program is the large amounts of cash you begin amassing almost immediately. A 1987 survey by the Association of American Geographers (AAG) of all North American geography departments cites $28,230 as the average annual salary for assistant professors in American geography departments, and $36,552 (in Canadian dollars) for assistant professors in Canadian geography departments (AAG 1987). That amount of money makes it almost possible to live in some metropolitan areas. But don't get me wrong—it *is* a salary! Your expectations rise far out of proportion to the real increase in wages over whatever compensation you received as an indentured servant (a.k.a. teaching or research assistant). As impressive as it initially seems, that paycheck will have to be more elastic than the definition of academic tenure to keep you living in a manner to which you would like to become accustomed. Stupendously consumptive eating and drinking habits commence right about the time the first check arrives. Trips to the store find you weighing the possible merits of boutique brews and different Tennessee sour mashes for the first time in your life, and contemplating salmon steaks and upscale broccoli as rewards for your new status. You actually consider buying those expensive books and bookshelves you've always wanted. And of course there are those old loan debts coming toward you, snarling and drooling, as your six-month grace period evaporates. Under such circumstances, who cares about tending to the gingivitis?

Those Pesky Little Squids

> I was now trying to get the better of the stupor that had come over me, and to collect my senses so as to see what was to be done (p. 140).

The single most important notion you have to avoid is the idea that the incessant knocking at your office door, those inane and irreverent inquiries, and the incredible amounts of time you are wasting, are all caused by those irrepressible students. The stress you feel is *not* caused by students. These *people*—remember that term—are merely a material manifestation, albeit a compelling one, of your new station in life. It is true, however, that you will garner no promotional benefits for keeping your office door wide open at all times, for answering the phone every time it rings—even at your home on Sunday morning!—or for amassing legions of "advisees."

Two characteristics of contemporary undergraduate students immediately catch your attention. One feature option of many first-year students is the ability to be content, even proud, to attain a C in your course. Motivation is an elusive virtue. Because you applied yourself in college (well, most of the time anyway), even if the course subject matter wasn't especially thrilling, you think all students are generally disposed to do the same. Such is definitely not the case. You may catch some of their reactions as you post final grades: Some become prayerful at receiving a D, and many positively levitate upon seeing a C next to their I.D. number. Now, you may accept this from business or engineering students who simply are not motivated to learn geography, but what will you think once you realize that half of your geography majors are likewise content merely to pass the course?

The second characteristic is that students seem to alternate between visions of the professor as soothsayer, as the dispenser of "truth"—"don't make me think, just tell me what I need to know for the exam" and "what is the *best* classification scheme for a choropleth map?"—and the professor as linebacker, standing between them and their blind job goal, ready to throw a poor grade as a clothesline tackle—"My parents won't send me to law school unless I keep my GPA up, so I *neeeed* a good grade in this course."

Most of today's first-year students have no idea of what to expect from a professor and from institutions of higher learning. Nor do they know how to study, use a library, or think about and use words, numbers, and symbols. Numerous books appear telling us where the blame belongs. But the fact is that as a junior member of the faculty you are nagged by the twin thoughts that (a) there is very little you can do about it and (b) they are pulling you down with them. You think you feel your mind begin to mollify. It can be stimulating to see the excitement of a prospective geography recruit, but that happens so infrequently that you occasionally feel like leaving the boat and wading into those dark, surging waters to save the masses yourself. I've seen junior faculty drown that way. Our universities have become day care centers for the vocational training of post-high school adolescents. Don't worry, it's only the decline of Western civilization. Watch and take pictures as we all spin downward.

If Not For the Courage of the Fearless Crew the *Minnow* Would Be Lost

> In the meantime the breeze that had headed us off fell away and we were dead becalmed, drifting about in every direction (p. 139).

Geographers could make great faculty colleagues, because the "circle the wagons" mentality that pervades most geography programs disallows the more intense displays of egomania and condescension that might otherwise be common. Nevertheless, there is a pattern to the behavior of senior faculty in geography (with a few scattered exceptions that you may encounter) that is emblematic of university faculty in general and so different from your graduate school experience that it warrants mention.

Faculty members often think it is to their advantage to divide into factions and join the battle for self-interests and for even the smallest bits of "turf." (I am reminded of the Uncle Vanya character in Woody Allen's movie, *Love and Death*. He was stark raving mad in his later years, wandering incessantly with a small mound of dirt in his hands that he would show to anyone who cared to see it. On this mound were a toy house, tiny bits of shrubbery, and so on. Allen's character narrates to the effect that, "That's Uncle Vanya. All he ever wanted was to own a small piece of land.") Your application, interview, and eventual hiring were undoubtedly analyzed in the context of turf battles, with faculty members asking themselves, "What can this person do for me?" You got the job, but at least one person on the faculty, and perhaps nearly half, opposed hiring you and may want as little contact with you as possible because your presence symbolizes a victory for the "other side." Soon, some may try to demonstrate their ability to wield power by placing you in awkward situations. Others will not openly be manipulative, but instead might believe that you are the program's new savior or technical support monkey. Still others will rise to your defense, if necessary, because it is politically opportune for them to do so. Regardless of the specific situation, faculty members will continue to have preconceived notions of who you are, what kind of research you do, and why you were hired—none of which may coincide with reality.

Self-reliance and a concern for self-interest are required of the occupants of any institution. Nevertheless, it is worthwhile to emphasize that no one has the interests of junior faculty in mind when decisions are made or duties handed out, least of all the senior faculty. Demands on your time accumulate at such a rate that it becomes difficult to sort wheat from chaff, especially when most tasks take you by surprise and all come with "due yesterday" stamped on them. Saying no occasionally is medically advisable but professionally dangerous, because votes on your tenure at the departmental level can hang in the balance.

An atmosphere of anti-intellectualism often hangs in academic corridors like a pall over a becalmed ship. To the extent that faculty members actually address each other, their dialogue consists of commiseration rather than an intellectual exchange of ideas. This will be apparent across campus and is not restricted to geography. "Collegiality" usually means a hello, idle chatter, or perhaps an ear for a complaint that needs airing. Otherwise, there is usually no time, patience, or inclination for discussion.

Observing the Crew in Action

> Scarcely had I secured myself in my new position, when we gave a wild lurch to starboard, and rushed headlong into the abyss (p. 144).

Departmental faculty meetings can at worst be a source of idle amusement for junior faculty and at best an unending source of information joyfully acquired by the intent observer. Faculty meetings have approximately the same therapeutic value as the World Wrestling Federation: they provide a catharsis for both spectator and performer caught in the academic maelström. The rooms where these exercises take place function as colosseum, sanitorium, and centrifuge, where the anti-cyclonic forces provide a perfect, if temporary, countervalence for the ensnaring cyclone outside.

As a personality- and location-dependent phenomenon, these gathering-so-we-can-divide soirees undoubtedly differ from campus to campus. Simply thinking of our own departmental staff meetings can still bring me out of the deepest of doldrums. The opportunity to *witness* the attempted collegial prosecution of departmental business far outweighs whatever real administering accidentally gets accomplished.

Placing geographers of different topical and methodological persuasions in close proximity promotes the possibility of confrontation as well as the primal urge to conceal information and personal agendas, or to mask them in unrelated jargon. Individual agendas quickly lead to empire building and turf protection, ultimately leading to separate, rather than collective, goals. The intriguing thing is that faculty everywhere seem to love this infighting.

I cannot overemphasize the importance that "deep" observation, mental notetaking, and behavioral interpretation at faculty meetings can have for junior faculty. This is the context within which you have to operate and within which your own developing attitudes and behavior will be evaluated. At least for the first year, you should stay out of departmental politics and avoid volunteering for any committees or administrative assignments, unless you know the precise workload involved and, more importantly, the symbolic context within which your actions are likely to be viewed. The better, more rewarding role is that of the "watcher": unbiased, alert, thoughtful, but not distant. It is a chance to accumulate a measure of political savvy as you observe and mentally assemble the different "camps," and, for example, try to determine the forces lying behind the decision to hire you.

Who's Got Their Hand on the Tiller (And Why)?

Developing a sound working relationship with the department chairperson is very important, especially if he or she actually seems to have a grasp of the situation in the department. Probably the most important aspect of this relationship is personality. The chairperson's perception of your needs and decision about meeting those needs in timely fashion are governed to a great degree by your mutual appreciation of each other's attitudes, motives, and preference for Monk or Mozart.

The newcomer-chairperson relationship takes on added significance when you hold a geography post in a department that includes other disciplines, and especially if geography does not dominate. Here, chances are good your chairperson will not be a geographer, will never have had a geography course, and will not know much about the subject matter and research methods of the discipline. Yet, you will have to justify repeatedly your course content and financial expenditures because that is the channel through which your funds flow. Material research needs that are usually met at the departmental level will require continual, detailed explanation, which you will grow tired of providing, such as, "No, physical geography is not merely the study of landforms."

The amount of paperwork and administrative committee work hardly varies with the size of a department. But the energy and time siphoned off by this type of "service" increase when geography is not only a small program, but one housed in a nongeography department. Under these conditions, geographers tend to construe programmatic issues as signifying either an impending cataclysm or a free pass to nirvana.

At my university, geography was the runt of a starving litter that included public administration, international relations, government and politics, and a couple semi-autonomous "centers." One year we were conducting five faculty searches, each in a different field. In my third year, I chaired the geography search as a way of *reducing* the extraneous claims on my time. From the chairperson's perspective, I was one of many faculty with divergent interests and needs, all of whom were competing for pieces of the same small budgetary pie. We were very lucky to have an understanding chairperson, but we still battled constantly for attention. It was hard to secure basic funding for essential cartographic supplies, let alone for fundamentals like wall maps, when more than 75 percent of the department faculty got by quite nicely on paper clips and legal pads. From the chairperson's perspective, it was hard to reconcile 20 percent of the faculty—five out of twenty-five—in a program carrying only 5 percent of the department's majors—fifty-five out of approximately 1,100—with the fact that we were demanding more than one-third of the department's budget. After many years of constant effort by *all* members of the geography faculty, punctuated by periods of pronounced disappointment, we have succeeded finally in gaining departmental status and in slightly increasing our staff.

You should get to know your college dean, as well as selected, relevant "deanlets." Learn the administration's image of geography. Developing a standard line of patter about your particular field of study, and about geography as a whole, that does not condescend to the uninitiated, is another time-consuming but worthwhile endeavor. It is far better to be prepared than not. Learning the substance of your colleagues' research areas and their view of geography is essential, in order to present your program as an understandable, coherent whole. Otherwise, deans tend to wonder why they have a geography program in their college at all, let alone whether it should grow or, possibly, gain departmental status. Geography should avoid being perceived as the "braided stream" of the college, possessing numerous channels with little depth and only occasional random intersections—even if this is a true picture.

Judging the New Deck Hand

> Here the vast bed of the waters, seamed and scarred into a thousand conflicting channels, burst suddenly into phrensied convulsion—heaving, boiling, hissing—gyrating in gigantic and innumerable vortices, and all whirling and plunging on to the eastward with a rapidity which water never elsewhere assumes, except in precipitous descents (pp. 133–34).

In the years immediately preceding tenure evaluation, tension mounts alarmingly. However, even among newcomers, a darkness prevails in the tone of most junior faculty when asked about the tenure process. Without sounding overly critical, I think it fair to say that university administrators are sometimes motivated by financial paranoia, political greed, and an exaggerated sense of their own importance. They tend to arm the university to the teeth with an array of legal watchdogs. Consequently, the college dean's conversation with you about requirements for tenure amounts to little more than a few mumbled phrases. Established "rules" for achieving tenure are kept purposefully ambiguous because lawsuits brought by embittered faculty members have become so common. For example, at my own institution the guidelines have been changed almost yearly since I arrived on campus four years ago. Currently, I am "guided" by three sets of ambiguous requirements, attributable to the three different college deans we have had in the last four years (six in the last nine).

Equally illuminating are deans' hackneyed comments that your first year, or perhaps two, should be consigned to lecture preparation. Implicit, though decidedly *un*explicit, in this message is the reassurance that some kind of "break" will be given on the research side of your ledger in recognition of your need to develop course materials right away. It doesn't really work out that way in practice, however, because research productivity (usually measured by the number of single-authored publications in journals using the double-blind review process) is expected to accelerate dramatically after the first two years, to *compensate* for *lack* of early productivity. Everyone recognizes that the early semesters must be given over to course development, but no one warns you of the penalty you will have to pay.

How much time and effort should you expend in the development of course material when your professional survival, and the extent to which personal goals can be met, depend primarily on research productivity? Consciously or not, you end up ranking your classes in order of their importance and interest to you, reserving the greatest effort in developing course material for the one you feel is the most crucial, your "bread and butter" course. You must do a solid, credible job in the others, but anything more at this time diverts your attention from writing. As you develop this line of thinking, you realize it is the only sane approach, the only one that gives you the possibility of ascending out of this maelström intellectually intact.

Probing the Turbid Seas

Several things are amazing about lecturing in a university classroom. You will be astounded by both your capacity and willingness to expound upon matters about which you know precious little. You will be stunned to hear yourself give answers that are dead wrong. And you will marvel at what your body, clothing, and classroom equipment start doing. It is as if everything takes on a life of its own. Pacing proceeds at a velocity sure to put you through the wall before lecture's end, nervous twitches threaten to have you mistaken for a mass murderer, and your hands and arms disconnect from your body, flying around like the Wallenda Family. Your tie—geographers are only allowed to wear the avant-garde variety—turns inside out to attack your shoulders, and your shirt leaves your pants to try to escape from the tie. Hose sag, or develop runs, hair falls, heels break, and earrings sling across the room. The chalk and eraser won't come when called, and the wall maps neither come down nor go back up upon request. (I once gave a thoroughly unplanned "lecture" on Africa because a map refused to retreat and roll up into the rack.) If you are the stolid type, ten small indentations begin to appear and deepen at the edges of the lectern by the third day.

In the first two years, classroom teaching is more of an emotional experience than an intellectual one. (As time passes I suspect it becomes neither one for most faculty, another sad comment on the condition of higher education in North America.) In the first few semesters you are likely to have several out-of-body experiences, either because you are not sure who just delivered that incredibly boring and inaccurate lecture, or because you are so much in your element and at peace that ethereal forces elevate your soul above the pastoral tableau spread below.

Finally, you will wish for an out-of-body (or out-of-the-room) experience on that odd day when you waltz into class having no shred of a clue as to what you are going to say. Maybe you had a lecture ready but your five-year-old ate the notes at breakfast. Or maybe you merely forgot you had a class at that hour. It happens, and you can expect it to happen to you.

Who Designed This Vessel?

Typically, incoming junior faculty are expected to teach five or six different courses the first two semesters. Some universities and many colleges will require four courses per semester, in which case you can expect to complete enough research to publish about one article every olympiad. At such schools, your dean should have informed you that teaching takes precedence over research in promotion and tenure evaluations. If the reverse is true, you will soon feel the angular velocity inside the vortex accelerate. Invest some time in looking to move on.

In large departments, generally found in major research institutions, you can expect to teach four or five course sections per year, but to have only three or four different preparations. Program size matters, too. Smaller programs often require each full-time faculty member to teach many different courses, despite the fact that research productivity is the ultimate measure of accomplishment. Look at a course list of some small geography programs and try to imagine how they handle their curriculum. Small programs in research institutions sometimes develop delusions of grandeur. Instead of emphasizing one or two geography subfields, the faculty have overdeveloped the curriculum (especially the degree requirements) and can become entrapped by it.

You are hired to teach in a certain subfield and, frequently, to teach one or two general introductory courses. You should feel free to propose changes in the specialty courses of which you are now in charge. New ideas, your own preferences, and the needs of students are all adequate reasons for change. You may encounter some resistance from conservative colleagues who have grown comfortable with the old equilibrium and see little reason to deviate from it. This may be a problem, especially if you are filling the shoes of a prominent figure now retired. However, reasonable faculty members will expect you to make your mark on the program's structure and will give you the leeway to do so. Linkages develop within the curriculum, so that changes to one course may have ramifications for others, and this may create ripple effects through the whole department. But, aside from this legitimate concern, preexisting conditions should not prevent change.

Mapping the Funnel

> I now began to watch, with a strange interest, the numerous things that floated in our company. I *must* have been delirious—for I even sought *amusement* in speculating upon the relative velocities of their several descents toward the foam below (p. 146).

You will do no research for at least your first year. Jot potential research topics onto note cards and file them away. Starting anything new right now causes the frustration quotient to skyrocket, and will either leave you hating the project and wondering why you started it or will result in a half-baked product that eventually gets lobbed onto the most dusty shelf. There it will sit with the other detritus that was shopped around to different journals, all of which said the same thing: "Good idea; needs further development." So, give it up. Besides, you will already have learned to mistrust the library. It doesn't have what you need, and the staff can't find it when the library does have it. The interlibrary loan office is staffed by a single incompetent, and you haven't any time for it all, anyway. Instead, ride the tide into the dissertation if at all possible. "The Beast" will cough up a publishable manuscript fairly easily. When it appears in print, your self-esteem goes up immeasurably, making it so much easier to grapple for control over the rest of your professional life. If the dissertation is a bad memory, you can pursue related research that doesn't require hours building and reading references, considering methods, and so on. Major research projects just aren't possible at this early stage, because the mental sharpness you attained in graduate school is being blunted seriously during these first two years. Synapses simply don't crackle like they used to, so you might as well wait for the ascent.

The Counterflow of Time

The burden of inheriting the Judeo-Christian concept of linear time can weigh heavily on you. We think time is unidirectional, ordinal, continuous, and only tenuously linked to place. Frequently, we think it is important to be first. It matters little where you are, as long as you are "progressing" and "getting ahead." This view can drive you deeper into the maelström.

You will probably perceive important events in the past as more sustaining than your experience of the present (unless you are in the midst of a serious discussion about geography with an attractive member of the opposite sex). Contentment with the present is difficult, because there is little in it to please you, while you salivate at the promise of a better future. Reality is easier to envision in the future than to accept for the present, unless "calendar shock" strikes, and you become frantic or incapacitated by the sheer mass of tasks awaiting your undivided attention.

Notions of progress to, or decline from, a better or "golden" age are implicit in your attitude. You come to think you are likely to spend the rest of your academic life trying to link yourself to a university geography program as good as or better than the one where you earned your Ph.D. But, by definition, there are not enough of these first-rate schools to house all geographers. You will have the nagging feeling that you have slipped back a step and must redouble your effort to catch up, keep pace, and move ahead. It could take years and leave you shorter of breath each day.

You grow impatient with friends and family who mean well, but who don't understand your situation. All you can do is ride out the whirl alone.

It turns out, however, that time is circular as well as linear. What goes around, comes around. Descent transforms to ascent.

> It was not a new terror that thus affected me, but the dawn of a more exciting *hope*. This hope arose partly from memory, and partly from present observation (p. 146).

ASCENT TO THE SURFACE!

> The slope of the sides of the vast funnel became momently less and less steep. The gyrations of the whirl grew, gradually, less and less violent. By degrees, the froth and the rainbow disappeared, and the bottom of the gulf seemed slowly to uprise. The sky was clear, the winds had gone down, and the full moon was setting radiantly in the west, when I found myself on the surface of the ocean (p. 148).

At some point in your voyage, light appears, the vortex loosens, and you slowly rise to more tranquil waters. How this happens and at what point the change occurs are matters that still mystify me.

The novelty of a paycheck begins to wear off, and you settle into your new lifestyle. You gain control of your courses and learn how to teach geography. You begin to enjoy working with upper-division geography majors and graduate students. Before-class preparation no longer takes very much time. You have made the necessary adjustments to institutional realities. Notions of excellence in teaching remain high in most contexts, but must inevitably accommodate reality in others. Have you adjusted and "settled in," or have you compromised your standards? Only cold, hard self-examination may tell. The quality of students has not changed, but your demeanor has. Lightening the load of tension makes the ascent easier.

The department reaches a new equilibrium as your presence is now enmeshed with that of your colleagues. You develop an important bond with one or two, and an understanding with a few more. You agree to disagree with others, and simply avoid some. You learn to accept these relationships and seek intellectual stimulation accordingly. Establishing contact by telephone or correspondence with distant colleagues is more important for stimulation and personal growth in this respect. You should make time to write letters. Few departments are willing to allow unlimited long-distance calls. More importantly, writing is a better, more considered way to find out what you are really thinking about. Geographers seem reluctant as correspondents, but our relatively small discipline is in special need of such direct linkages.

You have come to terms with the structure of the department and the curriculum. You have gained a feel for the administration, and especially its image of geography. You are able to make a satisfactory start on your research program; a small publication victory is sufficient, while two are admirable. Above all, you have adjusted to the pressures associated with the culture's prevailing conception of time, and have either removed extraneous and distractive forces from your life or perhaps rejected the pressures outright in exchange for an alternative lifestyle. Life is good again.

> Those who drew me on board were my old mates and daily companions—but they knew me no more than they would have known a traveller from the spirit-land. My hair, which had been raven black. . .was as white as you see it now. They say too that the whole expression of my countenance had changed (pp. 148–49).

Now you are ready to think about going up for tenure.

REFERENCES

Association of American Geographers. 1987. *1987 AAG survey of departments of geography in the U.S. and Canada*. Washington, D.C.: Association of American Geographers.

Auden, W. H., ed. 1950. *Edgar Allan Poe: Selected prose, poetry, and Eureka*. New York: Holt, Rinehart, and Winston.

Doing Geography: A Perspective on Geography in the Private Sector

Douglas B. Richardson

To a graduate student on the verge of becoming a professional, private-sector geographer, the differences between working as a geographer in the private sector and in a university setting must loom large. They certainly did for me ten years ago. Although the differences are significant, as I will discuss, I must say at the outset that the similarities are, in the end, far more important.

A prevalent misconception confronts me at every annual meeting of the Association of American Geographers (AAG)—and I still attend each year, for good reasons—that academia and the private sector are two nearly mutually exclusive worlds, separate universes that might even have totally different physical laws operating. The queries I receive from acquaintances and colleagues upon my arrival at an AAG annual meeting often resemble those that Marco Polo must have received upon his return from his long journey to distant and mysterious lands: "What is it like? How are you surviving? How amazing!"

The unexciting tale, however, is that conducting good geographic research in the private sector is pretty much the same as conducting good geographic research in the university setting. The exciting reality, on the other hand, is that there now exists a significant, growing, well-funded, and dynamic geographic research capability in the private sector in both the United States and Canada. This offers an exceptional opportunity for top students and promises well for the overall development and vitality of the discipline. The emergence of a strong private-sector research base in North American geography creates preconditions for the type of highly productive and creative interaction between private and university researchers that has traditionally characterized growth phases of other scientific disciplines, including chemistry, computer science, biochemistry, electronics, and geology.

The size and sophistication of the private-sector geographic research base in the United States and Canada now rival those of the universities. The geographic research that private-sector geographers are conducting today is at the cutting edge both of the discipline and of science. Many of the recent advances in both theory and implementation of the growth areas of geography—particularly geographic information systems (GIS), environmental modeling, and regional economic development theory—are now being developed largely in the private sector, and only later are picked up in academia. The sophisticated research tools employed by private-sector geographers more often than not surpass those of academic researchers. The task of generating new, original data is now largely the domain of nonacademic geographers, as they are increasingly the only geographers in a financial position to undertake the collection of such data as part of their research. And the issues addressed on a daily basis by private-sector geographers occupy a "central place" in the scientific, social, and policy debates of the nation and the world.

Geographic research in the private sector is thus emerging as one of the most vibrant and dynamic foci within the discipline. Today there is an excellent opportunity to define and establish a strong and synergistic partnership between the academic discipline and private-sector geography. Consolidating this position should be among the top priorities of geographers interested in moving the discipline forward and in establishing a strong research base for geography in the United States and Canada.

This chapter addresses several issues that may be of concern or interest to geography students or faculty members who are considering the transition from the university to the private sector. The discussion is drawn primarily from my own experiences, as I moved from the university setting to private research institutions in 1978, and then formed a consulting firm, GeoResearch, Inc., in 1980. GeoResearch's client base today is largely (approximately 85 percent) in the private sector, with the remainder being state, federal, or international government agencies. GeoResearch's private-sector clients are mostly large industrial or natural-resource-based companies, such as Burlington Northern Railroad, Basin Electric, Champion International, and Exxon. Most of our work takes place in the United States, while about 20 percent involves international projects. This essay therefore applies primarily to the U.S. context, with which I am most familiar; where possible, I draw parallels to the Canadian situation, with which I am much less familiar.

MAKING THE TRANSITION FROM THE UNIVERSITY TO THE PRIVATE SECTOR

What is it that private sector geographers do?

Private-sector geographers address the same range of research problems that university researchers study. Their daily work involves the same type of research activities: library research, database development, fieldwork, computer analysis, thinking, and a great deal of writing.

In my case, however—and I can only speak for myself here—the transition from graduate school to private-sector employment included some important changes, too. First, the pace and intensity of work accelerated dramatically. I felt as if the world suddenly had been shifted into a higher gear. Travel increased greatly and responsibilities grew quickly. Also, I had to interact with a much broader cross section of people than at the university. All this gave me the sensation of being stretched—emotionally, intellectually, and physically. My early research projects, on American Indian energy policy issues, also had urgent and tangible social, economic, and environmental importance; I felt exhilarated by the prospect of having real work to do that made a difference in the world.

Second, the scale of projects on which I was working began to increase rapidly, and the amount of money available to conduct research soon reached levels beyond my wildest imagination as a graduate student. Whereas my graduate school colleagues and I spent a great deal of time and effort trying to scratch together $600 for travel or $2,100 for necessary supplies and research materials, in the private sector I soon was able to command research project budgets of $700,000, $1.2 million, and $2 million. This difference in the amount of resources available for conducting research was for me a major feature of the transitional landscape between the university and the private sector.

For better or worse, the capacity to conduct original research is often greater in the private sector simply because of the resources available. Much of the research conducted in academia is unfortunately confined today by impossibly small budgets, which often constrain fieldwork and limit researchers to rehashing existing data, government-compiled statistics, library summaries, and so forth. Alternatively, for example, on GeoResearch's private-sector environmental modeling projects, instead of using spatially inadequate or poor quality existing data (as is frequently done), we generally can afford to undertake large-scale air quality and meteorological field monitoring programs designed to collect precisely the data needed. Extensive fieldwork, so important to geographic research, suddenly became accessible. Solid, thorough-going fieldwork has become a hallmark of GeoResearch's projects.

For these reasons, among others, I personally found the transition from graduate school to the private sector an enjoyable one. Private-sector research has provided me with an opportunity to explore freely both theory and practice in geography's best-equipped laboratory—the real world.

What other differences might the graduate student or restless faculty member contemplating a future in the private sector expect to encounter between the two workplaces? This may surprise many, but it has been my experience and observation that more writing is required in the private sector. Research reports tend to be longer and more comprehensive (perhaps less concise?) than articles designed for academic journals; deadlines tend to come faster, and simultaneous work on two or more projects is not uncommon. Publication for publication's sake in small-circulation, specialized journals is not emphasized heavily in the private sector.

Although fieldwork tends to play a more extensive role in the research process of the private sector than in the university, it is not because of any difference in method or emphasis, but simply because greater funds usually are available for this generally costly activity. It has been my experience that the use of computers and availability of hardware and software resources tend to be about comparable, and excellent, in both private-sector and university settings. Teaching, of course, is not a formal part of the day for most private researchers. Continued learning certainly is, however, and this is fostered by frequent informal meetings and seminars and by project team interaction, during which teaching and "knowledge transfer" skills are highly valued. Sponsorship of specialized workshops and training seminars for clients, as well as student internship programs and new employee training, also place a premium on teaching ability and experience.

Continuing education is likewise a fairly regular part of most private-sector geographers' lives. Travel to two or three professional meetings per year and attendance at specialized workshops are routine for professionals at our firm. Most private firms also provide time off and pay tuition expenses for continuing education classes. The greatest resource for continuing education available to private-sector geographers, however, is the experience and knowledge acquired through active participation in challenging project work on a daily basis.

The topics addressed by private-sector geographic researchers span the full breadth of the discipline. As an example, at GeoResearch, Inc., which focuses primarily on natural resource issues, research projects still involve a very wide range of topics. This seems endemic to geography, fortunately, whether private or academic. Our firm's current and recent research, for example, includes the following types of studies:

- Locational analyses of hazardous waste generators and regional disposal capability
- Site selection research for major facilities, including electrical generating stations, coal mines, and industrial plants
- Numerous studies relating to the proposed siting of the Superconducting Super Collider facility (perhaps the largest facility siting exercise ever carried out in North America)
- Analyses and mapping of complex mineral ownership patterns
- Regional air pollution studies
- Dozens of meteorological and pollutant dispersion modeling analyses
- Litigation resolution strategies regarding many environmental issues
- Groundwater pollution studies
- Research on GIS database structures
- Regional economic development studies
- Socio-economic impact analyses
- Land status, jurisdictional, and policy analyses relating to energy development on American Indian reservations
- Acid rain studies for Dutch, Italian, and American clients
- Agricultural resource studies involving custom designed remote monitoring technologies developed in-house
- Extensive modeling of rail-coal transport systems

Our firm's research projects have taken us from American Indian reservations to the Halls of Congress in Washington, D.C., and from the volcanoes and fumaroles of southern Italy to the synthetic fuel plants of Beulah, North Dakota. That there is a need for geographers who can *do* geography is clear. A dynamic, successful, and growing group of private-sector geographic researchers is beginning to meet that need and is demonstrating on a daily basis that geography has something exciting, current, and real to offer the world.

EDUCATING GEOGRAPHERS FOR WORK IN THE PRIVATE SECTOR

Geographers planning to work in the private sector should obtain the best and broadest education possible. Private-sector firms want the best educated people they can get. Work in the private sector is, if anything, more demanding than work in the university setting. It is a serious mistake for geographers intent upon gaining a leadership position in private industry to handicap themselves by opting for the so-called "applied geography" curricula cropping up in some departments. In my opinion, these narrowly focused curricula do a disservice to the student by undermining one of the basic strengths of geography, which is its breadth of focus. Although technique, particularly GIS, is important, a solid grounding in geographical theory and a diverse background are likely to matter far more to the later development and advancement of your career in the private sector. In the long run, technique, although important, is no substitute for substantive knowledge and well-honed analytical ability.

For top students contemplating a management or senior research career in private industry, I would suggest emphasizing the classic core strengths of geography. Develop a strong multidisciplinary background, while at the same time learning one substantive area of study thoroughly. Concentrate on developing writing skills and GIS skills, both of which will be essential to a successful career in geography in the private sector. I believe that the classical education of the geographer prepares students exceedingly well for the more broad-based responsibilities that come during the later stages in their careers. Technical skills may get you the initial job, but the breadth of your education and your ability to work creatively and effectively will be decisive factors in your career advancement, and in the long-term significance of your accomplishments.

A few structural differences between the university and private-sector work environments should also be mentioned. First, the private sector emphasizes ongoing performance over past academic credentials. There is no tenure in private industry. If you want to hold senior positions of responsibility, you have to perform well, and continue to do so. Work in the private sector involves a "state of doing" in addition to a "state of being." Second, it is important to be able to work with others. A great deal of work in academic institutions is carried on by individuals working alone. Students sometimes work together with peers on class projects but not often in closely integrated, hierarchical groups. Hierarchically arranged work teams, with many talented people at both senior and junior levels of expertise, are common in private research. The ability to work well in such an organization is important to success in private industry.

If the ability to work well together is necessary for success in the private sector, it is equally important that leadership in both the private and university research communities learn to work together and to establish a new partnership. This is essential if the transition from graduate school to the private sector by tomorrow's students is to reap its full rewards for geography in North America. Opportunities for such cooperation, as well as current obstacles to establishing these linkages, are discussed in the next section.

PRIVATE AND ACADEMIC LINKAGES: OPPORTUNITIES AND EXISTING OBSTACLES

Most exciting to me and, I hope, to students who may consider a future in geographic research, are the prospects for revitalizing geographic research in the

United States and Canada through the synergy that can be achieved by an improved integration of private and university research in geography. Yet it is unlikely that such a development can rest upon present notions of a grand distinction between "applied" and "nonapplied" geography, which characterize the rhetoric of the discipline and its organizations. Integration of the discipline can rest even less upon concepts of separate and noninteracting geographical theory and geographical practice. The widespread misuse in the discipline of terminology relating to a distinctive "applied" geography has led to an unfortunate notion that applied and nonapplied geographical research are somehow fundamentally different.

Worse, particularly for the progress of the discipline, is the derivative notion that the development of geographic theory and geographical practice involves separate, mutually exclusive activities. This notion suggests that the university, in the abstract, generates knowledge, which the private sector then applies. Recent presidential columns in the *AAG Newsletter* typify the myopia that persists in the institutions of the discipline regarding the relationships that are possible between university and private research (e.g., Jordan 1988, 1).

This naive worldview characterizes few other major disciplines, most of which have mature and powerful private research components that operate at the cutting edge of the science (e.g., chemistry, computer science, geology, and others). It is particularly ironic that academic geography has largely overlooked or deprecated the potential significance of a private-sector research base to the progress of the discipline, for the proper research laboratory of geography is in fact the real world. Geographic theory can only be tested by experimentation in geography's laboratory, and the data and results achieved in "real world" research create the only solid basis for new theory.

Moreover, the subject matter itself of geographic research is particularly accessible to the private-sector geographic researcher, who is directly conversant in the ways of the daily world. As David Lowenthal points out, "The subject matter of geography approximates the world of general discourse; the palpable present, the everyday life of man on earth, is seldom far from our professional concerns." Lowenthal further notes that "geography observes and analyzes aspects of the milieu on the scale and in the categories that they are usually apprehended in everyday life" (Lowenthal 1961, 241). Geographic research is privileged to be able to start with familiar relationships and phenomena. More than in any other field, the focus of geographic research is fundamentally congruent with the objects and structures of daily life (Richardson and Stephenson 1977). This is a powerful unifying dynamic that can be used greatly to the advantage of our discipline if we resist the temptation to shroud the familiar in pretense. It is clear that many current divisions within the discipline between university and private-sector workplace are largely unnecessary, justified neither by the nature of our research nor by the discipline's broad traditions.

One important step toward the successful integration of private and academic research capabilities in geography would be to focus our attention more on the substantive nature of our geographic work and the research problems and challenges it presents, rather than upon place of employment or upon generally meaningless distinctions between applied geography and nonapplied geography. Misconceptions regarding this terminology and its confusion with place of employment already have created artificial and counterproductive divisions within the community of geographers.

A further extension of this confusion for the student considering the private-sector workplace has been the recent creation of an academic bureaucracy whose new specialty is the so-called *separate field* of applied geography. This segregation of applied geography as a specialty subfield (without a common substantive thread), fosters a sense of applied geography as somehow a separate category of geography, requiring both faculty whose specialty is a technique-oriented applied geography and a separate curriculum track for the student. Few in chemistry, however, would suggest that the chemist who works in a laboratory at, say, DuPont, should have studied a fundamentally different type of chemistry from one who would work in a laboratory at the university.

As a private-sector geographer, it is not surprising to me that I generally have found more common ground and mutual interest with solid academic researchers in my field than with any of the various subgroups established to foster applied geography. Without a focus on substance, such groups, in my experience, tend to become dominated by turf-oriented, bureaucratically inclined individuals. They quickly lose the vitality that comes with real, substantive sharing of research interests.

When the emphasis shifts away from place of employment to actual research topics, however, excellent rapport generally follows between academic geographers and practicing geographic researchers in the private or nonuniversity public sector. Both groups, essentially, are doing much the same thing: bringing geographic methods and theory to bear on real research problems. This is a unifying activity. The similarity of doing real research in geography and the mutual interests and problems it generates should naturally bring the best workers together, whether they work for a university or for a private research firm (Abler 1988).

Thus, what is required is not further "recognition" of applied geography, nor the erection of institutional structures within the AAG or the Canadian Association of Geographers to serve these somehow "special" needs, but, rather, integration of private and university researchers around common research interests. The growing vitality of private-sector geographic research offers one of the best means on the horizon for rejuvenating academic geography and the discipline of geography in general. With salaries and research budgets in the private sector more than competitive with those available in universities, we can expect that increasing numbers of the best young research geographers will opt for positions in the private sector. It is important that *integrative* linkages, based on shared substantive interests between university and private-sector researchers, rather than *segregative* structures, be established if the opportunities for utilizing private research resources for the advancement of geography in North America are to be realized.

Geographers working outside the university, also, have sometimes contributed to the present divisions within the discipline. For nearly a decade, the Applied Geography Specialty Group of the AAG has belabored the issue of the organizational relationship of academic geography and nonacademic geographers to the institutional structures of the discipline. Some members of the group still adopt a combative attitude toward academic geography and seem to feel that a significant mission of the group should be to develop institutional strategies to breach the ramparts of the discipline—to somehow lodge the Trojan Horse of applied geography deep within the AAG. However, a glance at recent AAG annual meeting programs shows that private-sector and nonacademic public-sector research is already percolating very nicely throughout the discipline, more by force of constructive example than by institutional design.

The greatest progress for the discipline will occur by setting the example of good geographic research—whether applied or without application, whether conducted by employees of universities or of private firms—and by fostering collegial rapport and information exchange among all of the members of our discipline. Past contentious obsession with the institutional position of private-sector geography with regard to the academic sector of the discipline should shift to a more constructive focus on simply "doing good geography" and establishing professional rapport within our own research areas and throughout our national associations. The influence and status of geography in North America ultimately will rest more heavily on the success of our own research activities than on any brittle institutional structures erected within the organizations of the discipline.

Informal efforts aimed at fostering integration of the applied and nonapplied segments of the discipline within its formal institutional activities should be a primary concern. These should be cultivated in several ways: joint research projects; two-way consulting between university and private researchers; incorporation of top university faculty on the boards of directors of private research firms; use of top private researchers as adjunct faculty or for special seminars in universities; fostering of internship programs; joint development of proposals for large-scale government and international research projects.

Many practitioners have suggested that geography needs to develop an array of promotional efforts both within and outside the discipline to raise the visibility of geography and to publicize the value of geographic skills and knowledge to the outside world. Although such efforts may be helpful in the short run, their ultimate success depends almost wholly on the intrinsic worth of what is being promoted.

I suggest, instead, that our best advertisement will be our own good example, and that our primary strength lies in the fact and vitality of our productivity and in the work that we are accomplishing. Because private-sector geographers are actively *doing* geography, the private sector is on the growing edge of the discipline. Therefore, it will continue to attract some of the best and most capable of those students inclined toward geography. Establishing an effective working partnership between the growing private-sector research base and academic geography offers clear benefits to both sectors, to the progress of research in the discipline as a whole, and to today's and tomorrow's students who will be rolling back the frontiers of geographic research at private firms across the continent in the decades ahead.

I urge my colleagues in all workplaces of geography to worry less about our image or formal position within the discipline and to concentrate instead on developing those strengths we possess, which, in interaction with all other dimensions of geography, could well prove to be the engine that will get the discipline of geography moving again. To do this, it is not necessary that we waste great amounts of time and energy convincing others (even others within the discipline) that private-sector geography is "for real." It is usually more difficult to try to persuade others that a necessary goal is achievable than it is to go out and simply do it. We are fortunate today that much has been done already.

For today's student, there can be no better reason for studying geography in graduate school, and no better hope for geography in general, than the options provided by the diversity, vitality, and excellence of the private-sector geographical research that is being done today.

CONCLUSION

It has been my experience that good geographic research in the nonacademic setting requires the same methods, techniques, research approaches, and educational background as does good geographic research in the academic setting. In many cases, the standards are higher, and the resources available for the researcher much greater, in the private sector than in the current university setting. A university program designed for researchers "applying" geography should thus be at least as rich and as diverse as one designed for researchers "not applying" geography. Students who seek leading positions in the private sector should avoid specialized applied geography curricula that substitute technique for theory and substantive knowledge.

Private-sector geography offers excellent research opportunities for top students in the decade ahead. The development of a strong private-sector research base for geography in the United States and Canada, as already exists in most other sciences, is at hand. Integration of private-sector and university research activity can create a powerful partnership for growth and revitalization of the discipline.

REFERENCES

Abler, R. F. 1988. What shall we say? To whom shall we speak? *Annals of the Association of American Geographers* 77:511–24.

Jordan, T. 1988. The intellectual core. *AAG Newsletter* 23(5):1.

Lowenthal, D. 1961. Geography, experience, and imagination: Towards a geographical epistemology. *Annals of the Association of American Geographers* 51:241–60.

Richardson, D., and Stephenson, D. E. 1977. Bridging the gap: The geographer's world is everyone's world. In *Applications of Geographic Research,* ed. H. A. Winters and M. K. Winters, 155–62. East Lansing, MI: Michigan State University.

Geographers and Government Employment

Barney Warf

Professional geography in North America has long been and still remains a discipline heavily oriented toward academia: its journals are filled with papers written mostly by and for academics; its conventions emphasize topics of academic orientation; and its priorities, reputation, and constituency largely reflect the interests of professors and students. This fact is not itself unhealthy, for geography, like every field of knowledge, needs to ensure that scholarly research is conducted and that as a discipline it is reproduced among successive generations.

There are dangers, however, in geography's dependency upon academia for its members, resources, and inspiration. The academic labor market has not grown much in the last decade, nor will it in the near future. Further, the overrepresentation of academics in the discipline may lead to a narrow perspective on the world and exclude opportunities to expand the field into new domains of research and employment. Therefore, the discipline of geography must take seriously the actual and potential role of its nonacademic members.

By most geographers, government employment has been considered, at least implicitly, a second-best alternative, a stepping stone toward bigger and better opportunities. Unsurprisingly, given this view, government-employed geographers constitute a small minority of the field. In 1984, they totaled only 523 persons, or roughly 10 percent of the membership of the Association of American Geographers (AAG) (Andrews and Moy 1986, 407). The perception that government employment has a second-class status—which is generally unspoken and unwritten but widespread nonetheless—is doubly unfortunate. It deprives the public sector of the skills and expertise that geographers have to offer, while it simultaneously deprives the discipline of the wealth of experience and insight to be gained by working for federal, state, and local public agencies.

This essay does not aspire to be a comprehensive review of all the jobs in government available to geographers; nor is it an exhortation to abandon the ivory tower for public service. Rather, it sketches some of the major types of government employment American (and, to a lesser extent, Canadian) geographers may wish to consider, offers some comments on the nature of the planning process that characterizes most jobs in the public sector, and then identifies some of the differences between government employment and academia. Finally, it suggests survival skills that are helpful when working in large bureaucracies. Geographers can, and occasionally do, succeed in public sector settings, but they are underrepresented when compared to economists and planners. The literature on geography and public policy has been silent on the nature of the work experiences involved, what everyday life is like, and the criteria for success—all of which influence personal decisions of whether or not to work in government. The object here, then, is to demystify the nature of government employment by offering interested geographers a realistic assessment of its strengths and pitfalls.

GOVERNMENT EMPLOYMENT OF GEOGRAPHERS

Geographers who work for public agencies are likely to find themselves classified together with those toiling in private industry under the label of "applied" geographers, a term generally taken to mean those engaged in nonacademic work. The term applied in this context is not a particularly accurate one. It not only implies that academics fail to engage in applied research, but it also lumps together geographers working in government and business as if they had identical professional concerns, when in fact the two groups often differ considerably. Different groups of applied geographers are similar, however, by virtue of their marginalized status within the discipline. Geographers work for governments for reasons that extend beyond the fact that governments demand the services they offer; they are often motivated by a desire to serve the public interest, and a belief that governments perform socially necessary functions that private industry does not—e.g., environmental protection.

Geographers working for government agencies perform a number of different jobs, including resource analysis, water and land use planning, cartography, remote sensing, studies of health care delivery, analyses of transportation problems, and others. They are rarely, if ever, called geographers as such; rather, they have job titles such as "planner," "analyst," or even "economist." They most often work in highly interdisciplinary settings and perform tasks that may or may not fall within the traditional domain of geography. As J.R. Anderson (1979, 269) notes, "Geographers working in government generally must have a pragmatic outlook that enables them to do the tasks assigned regardless of how geographical those activities may be." This means that government geographers occasionally find themselves working on a topic about which they know very little. Nongeographical work is especially likely for those entering with a bachelor's degree; those with advanced degrees are more apt to find specialized positions that fit their training more precisely. For most government jobs, a master's degree is sufficient, although anyone seeking higher-level, analytical positions would benefit from a Ph.D.

Most government-employed geographers work at the federal level (see Table 1). In the United States, roughly one-third of all applied geographers work in the Washington, D.C., metropolitan area (Monte 1983; Smith and Hiltner 1983). Geographers in the federal government have a distinguished history, including positions such as Geographer for the State Department, Geographer for the Defense Mapping Agency (DMA), and the Chief Geographer for the U.S. Geological Survey (USGS) (J.R. Anderson 1979; Wilbanks 1985).

Opportunities for geographers are found in several federal agencies. The DMA, for example, is the nation's largest employer of cartographers, and the USGS uses many cartographers for its topographic map series. The Census Bureau is particularly important, not only because it generates the most commonly used sources of social science data, but also because it employs a variety of analysts for survey design and data analysis. Other positions in urban or regional analysis may be found with the Bureau of Labor Statistics, the Bureau of Economic Analysis, the Economic Development Administration, the Interstate Commerce Commission, the Agricultural Research Service, and the Office of Naval Research. Physical and environmental geographers are often hired by such agencies as the Environmental Protection Agency, the National Oceanic and Atmospheric Administration, the Forest Service, the Army Corps of Engineers, the Bureau of Land Management, and the Soil Conservation Service. The requirement of Environmental Impact Statements for newly proposed projects has significantly increased the demand for geographical skills for the study of environmental topics and to assess economic impacts, such as traffic congestion. This is true not only within government agencies but also among numerous consulting firms. Additionally, the federal government is the largest provider of contracts to geographers engaged in consulting (Stutz 1980).

Public-sector employment is relatively insulated from the effects of the business cycle and is hence more secure, although generally lower paying, than private-sector employment. However, it can be influenced by oscillations in the prevailing political climate. The federal budgets of the 1980s, for example, have limited employment in several areas that typically employ geographers, such as the Census Bureau. Whether the federal job market will improve in the future is an open question. Clearly, government service is of less significance to geographers in the United States than in many other industrialized nations, where geography is often synonymous with planning.

While most geographers in the public sector hold jobs at the federal level, the most rapid growth has occurred in state and local governments, where roughly 40 percent of government-employed geographers now work. In the 1980s, as federal hiring levels have tapered off, state and municipal governments have become

Table 1
Distribution of government-employed geographers

	1967		1978		1984	
	Number	Percent	Number	Percent	Number	Percent
Federal	291	72.5	248	51.5	310	59.3
State	34	8.4	73	15.2		
Local	50	12.5	115	23.9	213	40.7
Regional agencies	26	6.6	45	9.3		
Total	401	100.0	481	100.0	523	100.0

Source: Smith and Hiltner 1983; Andrews and Moy 1986.

increasingly important sources of employment. The growth of local- and regional-level employment of geographers has coincided with their dispersal away from the Washington, D.C., area (Smith and Hiltner 1983).

Opportunities for geographers exist in many state, county, and municipal governments. Most states and many cities, for example, have an office of economic development, and given the progressive internationalization of the North American economy, several have departments concerned with international trade, particularly export promotion. Other jobs may be found in state departments of natural resources, transportation, housing, labor, and commerce. City planning agencies are also an important source of employment, but there geographers must often compete with graduates of planning schools, who tend to be more familiar with specific issues about which few geography programs offer much preparation—e.g., zoning regulations. In addition, geographers may find positions in utility companies, school districts, port authorities, public utilities, and community boards.

While few geographers in the federal government address issues with a particularly strong relevance to specific areas, geographers in local governments are more likely to focus on regional topics, a subject about which they presumably can offer some expertise. A distinctive feature of local-level work is the relative lack of data, much of which is collected only at the state and national levels. As a result, although local issues are not necessarily less complex, they are generally more difficult than national ones to approach with methodological tools requiring considerable data.

In Canada, the opportunities for geographers in government to some extent mimic those found in the United States, although the heavier state intervention there offers relatively more employment opportunities. The discipline of geography tends to be much better known and respected north of the border. In addition to the various departments and ministries at the national level, geographers may find openings at such agencies as Statistics Canada, the National Harbours Board, the National Research Council, and the Economic Council of Canada. At the provincial and local levels, where most urban planning occurs, jobs in cartography, transportation analysis, population and economic forecasting, and housing policy are frequently available.

POLITICS AND THE PLANNING PROCESS

Because the bulk of what geographers do when they work for governments is related to planning in its broadest sense—including urban and environmental planning and the analysis necessary to formulate and implement plans—some attention should be paid to the nature of the planning process. In most highly industrialized nations, governments play a major role in the distribution of resources, amelioration of problems, and construction of landscapes, thereby deeply affecting the everyday lives of citizens (see Scott 1980; Clark and Dear 1984). Planning is central to this process, for it facilitates the smooth operation of government intervention and occasionally allows governments to anticipate problems before they arise (although this is not often the case). Thus, 95 percent of all professional planners work for governments. As a considerable body of literature attests, geography as a discipline often intersects with the planning process (Christensen 1977; Mattingly 1974; McNee 1970; Smith 1975; White 1972).

As planners of one type or another, geographers confront a wide array of questions, not all of which are easily amenable to analysis. Environmental geographers, for example, may research and propose answers to questions such as the following:

How can the rate of soil erosion in a particular area be stemmed?
What is the least expensive way to minimize salt water encroachment in the local wetlands?
What is the likelihood that a flood of a given magnitude will occur in an area?

Similarly, economic geographers may address issues such as:

How is the local labor force responding to rush-hour congestion at a particular bridge?
What is the potential demand for a telecommunications facility in a certain city?
How will the decline in the teenage population affect wage rates and hiring practices in the local retail trade sector?

Clearly, tackling questions such as these requires a broad range of skills and resources, and a knowledge of the planning process and the social context in which it occurs.

The planning process portrayed in the academic literature unfortunately bears little resemblance to the planning process in reality (Scott and Roweis 1977). For example, zoning in practice is less concerned with an abstract form of "social engineering" than with the determination of whether a particular building satisfies specific height and density requirements. What appears grand in theory is sometimes boring in practice. Similar chasms between appearance and reality arise in the perceived and actual nature of government employment.

One advantage in working for the government is that it provides an opportunity to observe bureaucratic decision making as it actually works. In practice, the planning process is rarely, if ever, the result of a systematic assessment of alternatives, hypothesis testing, or the optimization of some objective function, as asserted by the well-known but generally unrealistic "rational comprehensive" model of planning (see Faludi 1973; Burchell and Sternlieb 1978). Rather, planning is a considerably less formalized and more political process than that depicted by the orthodox interpretation (see J. E. Anderson 1979; Smit and Johnston 1983).

A frequent motivation for those seeking government employment is a desire to "serve the public interest," "make a contribution to society," or "find meaning in their work." It is, therefore, something of a shock for many to learn that even if the public interest can be articulated specifically, it is often not the highest priority of planning departments. At times, the interests of planning departments and the public may coincide. At other times, however, the academic's interest in unbiased research may run contrary to the interests of his or her superiors. For example, while the planner needs to make plans or analyses as specific as possible, managers and politicians often favor vagueness, so that trade-offs are obfuscated and controversy minimized. This does not mean that planning excludes socially meaningful work. Rather, planning in practice is invariably more complex than the embodiment of the ideals espoused in theory.

Planning is above all a political phenomenon, which means that the ways in which objectives and goals are set, resources allocated, information gathered and dispensed, jobs and promotions awarded, and methods employed reflect the efforts of managers and policymakers to advance their own agendas. This aspect of planning has received remarkably little scrutiny in the literature. In large bureaucracies, which generally exhibit rigid hierarchical structures, power is jealously guarded and "turfs" are closely protected. For example, different departments—even within the same organization—may withhold data from one another, refuse to work on the same projects, or needlessly duplicate projects to protect or advance certain interests. The circumvention of these problems requires an exquisite sense of diplomacy, wide-ranging personal networks, and a clear understanding of both the formal and "real" organizational structures of power and the division of labor.

Inevitably, the politics of work penetrate the contents of planning documents, motivating, constraining, and shaping the results, usually in rather subtle ways. Rarely are overt political interests spelled out so obviously that their presence can be easily detected on paper. Rather, it is what is *not* said, or *how* it is said, that is critical. Often the positive aspects of particular projects are emphasized—e.g., the number of jobs created—while the costs are downplayed—e.g., traffic congestion. Self-criticism of an agency is usually strictly discouraged by those who head it.

In government hierarchies, planning documents are frequently reviewed by a variety of persons involved with a given issue, many of whom have vested interests in forwarding a particular viewpoint. In practice, the review process generally means that analysts' work will be "sanitized" to conform with the interests of their superiors. Statements in documents that may offend some party or another—thus constituting political errors as important as any analytical ones—are likely to be excised. Even a highly polished piece may never see the light of day if publication would be politically inexpedient—i.e., likely to run against the claims of some parties involved, generate controversy, or make one department appear in a better light than another. Unlike academia, where it is the *product* that counts, in public bureaucracies the *process* of creating a document is at least as important as the outcome.

Within the government workplace, it is the senior management—which rarely includes geographers—who make the major policy decisions. Geographers in government, therefore, tend to work in advisory roles, making their contributions to the decision-making process after the fact—i.e., as substantiations of decisions that have already been made, often for political reasons. Unlike academics, planners and analysts tend to be detached from the products of their labors and have relatively little say about the application of their work in particular contexts. For example, the analysis of a project may be used as a public relations piece, or a forecast may be deliberately misinterpreted to emphasize a particular viewpoint. If mistakes occur, they are likely to be attributed to lower-ranking colleagues; a useful maxim to remember in these circumstances is "blame flows down, praise flows up." The people who manage public offices, agencies, and services are not omnipotent, but geographers working in government will find that their perspectives, styles, and personalities are critical in shaping the day-to-day environment in which geographers work, the levels of efficiency achieved, the priorities that are set, and the techniques that are utilized.

These forms of politics are by no means unique to public bureaucracies—they also occur in most large corporations—but geographers considering government employment must prepare themselves for the notably Byzantine structure characteristic of most large bureaucracies. In small towns, generally with less complex administrative structures, such politics are apt to be less pronounced; in large cities, in which every government office is enmeshed in a web of other offices and multiple levels of a hierarchy, they are likely to be more so.

All this is not to imply that work in the planning process is unchallenging or unimportant—on the contrary, often the most interesting issues arise in the analysis rather than in the policy decisions themselves. Rather, it is to acknowledge that even the most carefully designed plans may be set aside to gather dust if they do not conform to the prevailing political interests in the environment in which they are constructed.

SURVIVAL SKILLS IN GOVERNMENT EMPLOYMENT

What is the nature of everyday life at the workplace for the government-employed geographer? How can a geographer obtain and succeed in a public-sector job? While there are no firm answers to these questions, a few guidelines and caveats are offered here.

The geographer deciding between an academic and government career should be aware of the differences and the relative costs and benefits of each. Government employment has several advantages over a career in academia. In general, given the limited academic job opportunities in the foreseeable future, government employment offers a relatively wider range of employment options, flexibility in location, and usually higher salaries. Whereas academics often work in isolation, work in a government office generally involves more teamwork and interaction with colleagues. Government jobs offer freedom from "publish or perish" pressures, yet often open up opportunities for research and access to data, as the literature written by government-employed geographers attests. The trade-off, however, is that government-employed geographers must sacrifice some of their autonomy and control over how they spend their time: with one or more supervisors, one has relatively few opportunities to be self-directed. Government employment lacks the seasonal rhythms so prevalent in academic life, and offers few possibilities for sabbaticals or temporary stints elsewhere.

A few geographers manage to bridge the gap between academic and public-sector employment—e.g., by teaching part-time. Maintaining ties to academia, however, requires an exceptionally firm commitment, and the discipline has not made it easy. How are those working in the public sector supposed to feel, for example, when they read about the need for geographers to "protect their academic core"? Further, geographic literature generally fails to address issues closely related to applied work, and nonacademics often feel like outcasts at geography conventions.

Obtaining a government research or planning job is a function of skills, experience, and personal connections. For the manager faced with a stack of competing résumés, there are few substitutes for experience in the choice of a candidate to fill an open position. Graduating students are often confronted with the well-known paradox of having little experience and only a minimal chance of obtaining employment without experience. One route around this problem is for students to seek government internships actively. Internships quite often provide a "foot in the door" toward future full-time positions and give geographers a chance to investigate an agency before they make a substantial commitment to it. Many municipal government agencies offer paid summer positions. Applying for such positions, however, takes a considerable investment of time. Also, because public bureaucracies move very slowly, the process must be initiated several months in advance.

Often ignored in the popular job search literature is the importance of personal contacts. That is how most people find out about job openings. In addition, contacts weigh more than a little in the selection of a candidate, all else being equal. Many government positions are not widely advertised and are often designed to be filled through internal placement. It is essential, therefore, that young geographers maximize their contacts with persons working in government jobs; even a "friend of a friend" or an acquaintance of a professor may be a sufficiently close contact in this regard. This is not to say that personal contacts will guarantee a position, but it is an indication of the widespread and important role that informal networks of people and information play in the planning process.

Geography students considering government employment can prepare themselves by taking courses likely to have wide applicability in the future. Some of the most important abilities necessary to successful job performance are "generic" intellectual skills, particularly the ability to construct coherent arguments synthesizing different forms of information. All geographers must be able to write well, as much of planning work consists of preparing reports; unfortunately, style is often as important as content. All geographers should have a familiarity with computers, which does not necessarily mean a knowledge of programming, but rather the ability to learn word-processing and statistical packages readily. As personal computers have replaced mainframes in an increasing number of offices, their importance has risen steadily. All geographers aspiring to government service should take several courses in statistical techniques. Additionally, an intimate knowledge of data sources and their relative quality and accessibility gives one a tremendous advantage in any work environment. Most government-related research is highly empirical in nature, which can reveal the size of the gulf between abstract models and the earthy complexity of reality.

Beyond these general skills, courses should match the area of interest that students intend to pursue—e.g., soils, hydrology, or economic development. Because most government-employed geographers are engaged in work related to urban or regional analysis, a knowledge of regional models (such as input-output analysis) is valuable; several economics courses, particularly macroeconomics, should be mandatory. In many research positions a familiarity with finance is helpful, although it is a subject to which most geographers receive little exposure. Demography is also a widely employed discipline, so a knowledge of population projection techniques is valuable. Finally, geographers familiar with zoning regulations and transportation modeling methods will find themselves in a favorable employment position.

Managers and supervisors are generally unfamiliar with what geographers can do. While most planners include time as a central element in their work, it is surprising how infrequently they take space seriously—e.g., they overlook the necessity of mapping data, including transport costs in models, or using subregions to refine theories of broader processes. Not infrequently, geography is held to be synonymous with cartography.

Thus, a note to cartographers thinking of government positions is in order, as it is one of geography's most employable skills (as confirmed by Andrews and Moy 1986). The vast bulk of the cartographer's functions are now computerized, yet few geographers have much familiarity with available mapping packages. Conversely, office computer experts generally know little about cartography. Without aggressive marketing, however, computer cartographers are likely to find themselves reduced to electronic graphic artists, not analysts who incorporate maps as central components of a geographic perspective.

Given the nature of the public-sector work environment, it is important to note that the ways in which analysis on the job *should be* done and the ways it usually *is* done generally differ. In theory, of course, analysis proceeds slowly and carefully, with due regard to the assumptions and limitations imposed by the quality and availability of the data; in practice, this is generally not the case. The analyst or planner is usually under pressure to write a report in a hurry and rarely has sufficient time to gather the data necessary for a sophisticated analysis. In theory, elaborate models provide more accurate answers than does intuition; in practice, models that take much time to learn or implement cannot be used rapidly. Thus analysts are often forced by circumstances to employ rough, "back of the envelope" techniques relying upon intuition and anecdotal evidence, where an educated guess or approximation is the best possible answer. Such methods are not optimal in terms of their accuracy, but they are often favored by managers who have limited resources to devote to a particular question and are under pressure to come up with prompt answers.

Further, analysts may be limited by their supervisors' skills, which reflect their education, and when and where they went to school. For example, a manager may be more comfortable with demographic-driven forecasts than investment-driven econometric approaches, regardless of their applicability to certain situations, and may insist upon using the former or deny funding for the latter. Managers who have spent several years out of school are frequently unfamiliar with, and skeptical of, state-of-the-art methods they view as useful for university research but not in the "real world" (see Stutz 1980, 395). Many analysts working for governments find they need to simplify complex issues for their superiors, who often do not want to be bothered by details such as the methodology used to arrive at a given conclusion. All this is not to suggest that abstract models and complex theories are unimportant—indeed, they are often critical—but to realize that their implementation is deeply influenced by the particular social and political climate of a job.

Finally, a caveat for the geographer aspiring to a government research or planning position: good technical expertise and hard work are not enough to ensure success. The geographer who becomes an expert on, say, input-output analysis alone is likely to remain a technician with few other opportunities. The criteria for success are generally much broader than technical know-how, including "people skills" and a knowledge of office politics. These skills are particularly important in the public sector, where economic incentives to perform well are often minimized. Few things are more frustrating than to witness less technically qualified but more socially skilled persons receive a disproportionate share of the available job rewards. The rules governing such phenomena are unwritten and often unspoken—but crucial nonetheless—and rely heavily upon informal contacts. To advance their careers, geographers must learn how to work around bureaucratic inertia, cultivate personal ties, form alliances, know when to take the initiative, assess when to persist in one position and when to transfer to another, and so forth. There is little that students can do to prepare themselves in this regard except wait to acquire experience, although a course in organizational theory might help.

CONCLUSION

Whereas geography has given lip service to government employment as an integral part of the discipline, it has yet to acknowledge the full rewards (and costs) of working in the public sector. Because many governments are actively involved in addressing social problems at several spatial scales, employment in the public sector can offer geographers a valuable means of combining theory and practice and a way to minimize armchair speculation. Such experience can go a long way toward decreasing the insularity of academic geography and improving its popular image. Indeed, for academics, government employment can help to improve performance in the classroom: Stutz (1980, 393) notes that "it is essential for academic geographers to get real-world experience themselves before they can be adequate teachers of applied geography."

In the future, it will be increasingly important for geographers to avoid viewing their field in isolation from its engagement with government work. As theory is inseparable from practice, geography is likely to be changed by that interaction, finding its priorities, methods, and perspectives more oriented toward public policy. If the discipline fails to avail itself of this opportunity, it will forego a valuable means of strengthening its public image and social relevance.

If geographers are to take this lesson seriously, then they should take steps to encourage government employment. Geography departments, for example, can devote more resources to developing internship programs; alumni in planning agencies can be invited to speak to students, thus maximizing personal contacts; methodology courses can be oriented toward the techniques most commonly utilized by government analysts; professors can undertake consulting projects for local public agencies; conference sessions focusing on the needs of government-employed geographers can be organized; journals can publish more articles with a direct bearing on important planning issues. Many disciplines have fruitful interactions between members working in academia and those in the private and public sectors; there is no reason geography cannot be one of them.

Anderson, J. E. 1979. *Public policy-making*. New York: Holt, Rinehart and Winston.

Anderson, J. R. 1979. Geographers in government. *Professional Geographer* 31:265–70.

Andrews, A., and Moy, K. 1986. Women geographers in business and government: A survey. *Professional Geographer* 38:406–10.

Burchell, R., and Sternlieb, G., eds. 1978. *Planning theory in the 1980's*. New Brunswick, NJ: Rutgers University Center for Urban Policy Research.

Christensen, D. 1977. Geography and planning: Some perspectives. *Professional Geographer* 29:148–52.

Clark, G., and Dear, M. 1984. *State apparatus: Structures, languages of legitimacy*. Boston: Allen & Unwin.

Faludi, A., ed. 1973. *A reader in planning theory*. Oxford: Pergamon Press.

Harrison, J., and Larsen, R. 1977. Geography and planning: The need for an applied interface. *Professional Geographer* 29:142.

McNee, R. 1970. Regional planning, bureaucracy and geography. *Economic Geography* 46:190–98.

Mattingly, P. 1974. On the value of geography in planning practice. *Professional Geographer* 26:310–14.

Monte, J. 1983. The job market for geographers in the Washington, DC area. *Professional Geographer* 35:90–94.

Russell, J. 1983. Specialty fields of applied geographers. *Professional Geographer* 35:471–75.

Scott, A. 1980. *The urban land nexus and the state*. London: Pion.

———, and Roweis, S. 1977. Urban planning in theory and practice: A reappraisal. *Environment and Planning A* 9:1097–1119.

Smit, B., and Johnston, T. 1983. Public policy assessment: Evaluating objectives of resource policies. *Professional Geographer* 35:172–78.

Smith, B., and Hiltner, J. 1983. Where non-academic geographers are employed. *Professional Geographer* 35:210–13.

Smith, P. 1975. Geography and urban planning: Links and departures. *Canadian Geographer* 19:267–78.

Stutz, F. 1980. Applied geographical research for state and local governments: Problems and prospects. *Professional Geographer* 32:393–99.

White, G. 1972. Geography and public policy. *Professional Geographer* 24:101–4.

Wilbanks, T. 1985. Geography and public policy at the national scale. *Annals of the Association of American Geographers* 75:4–10.

ON WRITING

Conducting Research and Writing an Article in Physical Geography

David R. Butler

Publish or perish. You've heard it before, and you'll hear it again. Is it true? Yes, it is true in physical geography, at least from the perspective of academics. This essay is an attempt to share my several years of experience in publishing in a variety of journals, on a variety of topics in physical geography, and in both sole-authored and collaborative efforts. It is my hope that these experiences and the ideas that have developed from them will assist you (as a graduate student or recent Ph.D.) during the years in which you must learn to deal with the potentially frustrating, but enormously rewarding, efforts of publishing in physical geography.

WHY SHOULD YOU PUBLISH?

The issue of whether or not a faculty member should publish has changed dramatically in the past twenty years. In the late 1980s, there is no longer any debate about this: You simply must publish if you wish to be promoted and tenured in a university setting. The modern collegiate atmosphere and finances are such that the old days of "doesn't publish, but is a great teacher" have disappeared at virtually all universities with graduate programs; this attitude has even trickled down to undergraduate institutions. The reality of "publish or perish" has never been more true.

Beyond the issue of publishing for the sake of promotion and tenure, publishing is an extremely satisfying experience that doesn't become diluted with time. Seeing a paper develop from an idea, and nurturing it along through the long process of data collection, analysis, development of conclusions, and eventual writing is difficult and often tedious. But when the paper you have submitted is finally accepted, a sense of accomplishment unlike most experiences in your daily work environment will carry you to heights of good feeling. A sense of accomplishment, the knowledge that you have contributed to your discipline and perhaps to scientific advance in general, and a surge of pride are some of the major rewards of publishing.

Typically, physical geographers' initial publications come from their master's thesis and/or doctoral dissertation. These frequently provide the initial data sets and analyses necessary for publication. *Do not overlook your master's thesis as a source of publishable ideas and data.* It may hurt to reread your thesis, with its choppy writing style and naivete, but your master's committee would not have approved it if they hadn't had some belief in its scientific merit.

A paper from your master's thesis may not be *Annals* material, but so what? Try a state, regional, or "lower-ranked" national journal. The experience will prove good preparation for major publishing efforts to come and will illustrate to potential colleagues that you have the desire to become a productive scholar. And you may be surprised where you can publish the results of your thesis. *Every* student whose master's committee I have chaired or served on over a six-year period has had at least one paper result from the thesis, in journals as diverse in ranking and reputation as *Mountain Research and Development, Photogrammetric Engineering and Remote Sensing, International Journal of Remote Sensing, Northwest Geology, Geographical Bulletin,* and *Oklahoma Geology Notes.*

Your dissertation should serve as the major source of publishable material for the first few years beyond graduate school. Now, assume that all your dissertation results have been published. What do you do? On which topics do you conduct research now? My reply to this question would be the same as to graduate students searching for a thesis or dissertation topic: "Do only what truly interests you. Do not choose a topic simply because it is potentially fundable or currently 'in vogue'." After all, why do many people become academics? Presumably, because we enjoy the freedom and choices available in academia. Was there an aspect of your dissertation that you would have enjoyed devoting further attention to, but did not have time for during the doctoral program? Do it now! Did you write a dissertation that was funded by your major professor, when you really wanted to examine a different topic? Get started! Do you find yourself wishing you could go back and reexamine aspects of your master's thesis? Why not start today? I cannot tell you which research topics to pursue, but I will advise you to follow your inclination, and continue to pursue the topic as long as it makes you happy. If you are a physical geographer, what makes you happy will probably be inherently interesting and of sufficient scientific importance to result in further publications.

HOW OFTEN SHOULD YOU PUBLISH?

No one can tell you exactly how often you should publish during the early stages of your professional career, because everyone's desires and opportunities to publish will vary. Nevertheless, some simple guidelines do exist for the minimum level of expectation on the track toward promotion and tenure. Beyond the minimum acceptable level, it is up to the research program of each individual to decide publication frequency.

Brunn (1987) identified five types of research-oriented faculty:

1. The small number of individuals who publish papers in four or five major disciplinary and interdisciplinary journals each year.
2. A slightly larger number who are also very active in research and grantsmanship, but whose average annual volume of effort is less—between one and three major papers per year.
3. Those who are active and have ongoing research programs, but who may not attend annual meetings; these individuals may publish a paper every other year.
4. Those who have some research projects, but who rarely present their results at meetings and usually no longer publish papers.
5. Those whom Brunn termed "professionally inactive," who have no research program, do not attend meetings, and never publish in refereed journals.

As a young Ph.D. in physical geography, you cannot afford to be perceived as any lower than group-two status, or you will not be promoted. If you do not publish between one and three papers per year, every year, as an assistant professor, it is probably fair to say that you will be denied promotion and tenure. An unwritten rule for most major geography departments requires twelve refereed publications for promotion and tenure, the assumption being that you will average two papers per year over a six-year period before being considered for promotion. It is important, as Brunn (1987) points out, to be seen to be publishing consistently. A few years of high productivity, such as a group-one faculty member would achieve, followed by several years of relative inactivity at group-three or lower levels of productivity, would provide a signal that you have "burned out" or are not genuinely interested in being a consistent, continuing scholar.

If there must, therefore, be a rule of thumb for the frequency with which you publish, let it be a *minimum* of two papers per year for six years, with the majority of these in major national or international geography journals known to all your colleagues. (For every coauthored publication, increase the total by one if you are first author and by an average of one and one-half for papers in which you are not first author. Unfortunately, publications on which you are not first author are frequently dismissed by colleagues during the "numbers game.") You will want to publish in major specialty journals as well–and some of your research efforts will not warrant publication in top-ranked journals–but to be safe, publish at least twelve papers, with six to eight of them in major journals such as the *Annals,* the *Professional Geographer, Journal of Geography, Physical Geography,* and *Progress in Physical Geography*. And be sure that at least one of the papers is in either the *Annals* or the *Professional Geographer.*

WHERE DO YOU SEND YOUR MANUSCRIPT?

You (and your collaborators, if appropriate) wish to submit a manuscript for review and possible publication. Where do you send it? I strongly believe that you should always write a paper with a target journal in mind, with a backup journal also identified. Not to do so is, I think, naive and a possible waste of time. Why write a paper using the *Annals* editorial style and philosophy, if you plan on submitting the paper as a technical note to the *Journal of Glaciology,* the *Southwestern Naturalist,* or the *Climatological Bulletin*? You must learn how to target certain journals and how to adapt your writing style to journal specifications. Specific methods for doing so were suggested recently for the *Annals* and the *Professional Geographer* (Brunn 1987, 1988), and all physical geographers in the early phases of their careers will benefit from a close examination of Brunn's suggestions.

With that major proviso in place, how do you identify where to send your manuscript? You should know which journals are considered highly ranked, top-quality, national or international journals, and which are clearly secondary and tertiary outlets for papers of lesser or local significance. Several methods for ranking journals exist.[1] Your duty, as a new professional geographer, is to have a general idea which journals are highly ranked: (1) by the discipline as a whole; (2) by your peers in your specialized subfield of physical geography; and (3) by your faculty colleagues who will evaluate you for eventual promotion and tenure. Let's consider each category in turn.

1. You can easily get a feel for the top journals in your field simply by consulting articles that discuss journal rankings, by walking through your library, and by simple word of mouth from friends and associates. Certainly, we all know intuitively the reverence that most colleagues hold for the *Annals,* the *Professional Geographer,* and a few other major journals. To enhance your visibility in the field as a whole, you will wish to publish in this group (regardless of your own opinion about the quality of these journals compared with other geographic or more specialized journals). This visibility helps to establish the regional or national scholarly reputation that tenure and promotion committees look for.
2. You also need to build your reputation among members of your particular subfield in geomorphology, climatology, natural hazards, soils, biogeography, and so forth. These individuals frequently have a different perception of journal quality and ranking than do geographers in general (Turner 1988). Physical geographers are, I think, a contrary bunch: we think that many nongeography and interdisciplinary journals are more appropriate outlets for some of our most significant works. You will probably want to consider publishing some of your more narrowly focused, but very significant, research results in specialized, frequently nongeography, journals.
3. In a perfect world, you wouldn't have to worry about this, but you will because what your faculty colleagues think of your publication efforts will play a major role in how you are perceived in your department and how you are rewarded financially and by promotion. For a physical geographer—especially if you obtain a position in a relatively small department with few kindred colleagues, or where a majority of your colleagues fall within categories three, four, or five in Brunn's (1987) typology—it is desperately important that you learn which journals your faculty colleagues believe are the best. These are the journals in which you should publish.

[1]See Lee and Evans (1984, 1985a, 1985b), Peet (1985), Gaile (1985), and Kenzer (1988) for some of the controversy surrounding the question of ranking journals and citation analysis.

Do not assume that the high quality of your articles in specialized journals will be apparent. Your colleagues, in some cases, may publish only once every three to five years (Brunn 1987); they may never have published in a specialty journal or a nongeography journal. Also, remember that many of the so-called specialty journals, such as *Physical Geography, Earth Surface Processes and Landforms, Journal of Biogeography,* and *Journal of Climatology,* began publishing only ten or fifteen years ago. Your more sedate colleagues may not be familiar with them. You can also assume that some will not be familiar with *Quaternary Research, Arctic and Alpine Research, Zeitschrift für Geomorphologie, Monthly Weather Review, Review of Palaeobotany and Palynology, Catena, Ecology,* and similar nationally and internationally respected journals. Many young physical geographers who blithely believed that the quality of their work would be apparent to all have heard their colleagues say, "Nice job, but it's too bad that you can't get in the *Annals,* the *P.G.*, or the *Geographical Review.*" The belief will be that your work simply is not good enough for "the majors."

Learn where your faculty peers publish. Ask them for reprints of their works. They will probably be flattered and happy to provide them. These can give you some insight into which journals they publish in and approve of. Also, take the time to go through the last ten years of the *Annals,* the *Professional Geographer,* and other journals you are considering, to determine how frequently your colleagues utilize those journal outlets. For example, is *Physical Geography* a respected journal in your department? I examined all twenty-one issues of this journal (from its inception in 1980 till March 1988). I found seventeen articles, out of a total of 136, authored or coauthored by current University of Georgia geography faculty. Fully 12.5 percent of all articles in that journal can be attributed to faculty in my department, so I can conclude that in my department this is considered a very desirable outlet for publishing. Your department will differ, so take the time to check it out early in your new career.

HOW DO YOU WRITE THE PAPER?

Let us assume you have collected a body of data/observations, thoroughly analyzed it, and come to some conclusions you now wish to share with the professional community. Because writing one's first few papers is so different from the enormous task of writing a dissertation, it may be useful to describe the process step by step. I will use my own writing procedures as an example. The steps that follow are typical, although I do occasionally vary the format to a minor degree.[2]

I begin by identifying the journal to which I want to submit the manuscript; I usually have a backup journal in mind as well. It is important to identify the target journal in order to adopt an acceptable format for the paper. Be sure to get a copy of the journal's "Guidelines for Contributors," so you will be aware of the article format; the reference style required; the limitations in graphics style, size, and lettering; and the simple and mundane but important items such as required

[2]For some specific examples that reflect this process, see Butler (1986, 1987).

margins and the number of copies to be submitted. Editors have told me that one of the best signals you can send to the journal you choose is simply an indication of your concern for its style and format. Submit your manuscript in conformity with that journal's requirements. It goes without saying that the manuscript should be free of any typographical errors, omission of references cited, and so forth.

Once I have the journal specifications in mind, before I begin the actual writing I formulate a simple mental outline. (Whether or not the outline is committed to paper is an individual decision.) It usually contains most or all of the sections of a paper described by Brunn (1988): an introduction, background information/literature review, description of the study area (if the paper deals with a site-specific case), description and justification of the methodology employed, discussion of the analysis and results, and conclusions and discussions for future directions in related research. Be sure to consult the guidelines for contributors to ascertain whether the journal has any specific section titles, which kinds of headings and subheadings are appropriate, and whether the journal has any manuscript length, figure size, or text limitations.

During the outline stage, I like to envision which figures and tables (if any) the article will need for illustration and support of my ideas. I typically go ahead and format the data tables on a word processor, so I can see how they will actually look. Similarly, I have publication-quality graphics drawn up early in the conceptualization stage. I have my maps drawn, PMT figures created, and black-and-white photographs printed. Even if, in revision, I ultimately delete one or more of the maps or photographs, I have found that I will eventually use all such graphic materials. With graphics and tables complete, I can conceptualize more easily the flow of the writing as it encompasses and describes them. By all means, number the figures and tables sequentially in advance, and write out your figure captions before you write the paper. In this fashion, you will not forget why you are using these particular graphics or data, and their presence helps further solidify your own mental outline.

I also prefer to create a tentative list of references I will cite in the paper, before I begin to write. Again, using a word processor, it is easy to delete any leftover references you may not have used in your final draft, and having them in front of you during the writing stage will assist the ordering of your thoughts in your mental or written outline. Certain topics you write about will have a "must cite" list of significant papers and books, so why not have those references in front of you while writing, to ensure that you will not forget to include them when the "writing frenzy" hits?

When you begin writing your article, you may find the introduction very difficult, for this is where you have to provide the reasons why an editor and reviewers will want to read this paper, and what contributions it makes to the discipline (whether conceptual or methodological). The study area description is dry but easy to write, and becomes almost standardized if you work in a particular area for any period of time.[3] Description and justification of methodology are usually easy to write as an extended list with justifying references, *unless* the paper's main contribution is methodological. In that case, you will need to build a much stronger

[3]See Butler (1986, 1987) for standard study area descriptions from the two primary field areas on which I have published papers.

argument for using a new, and potentially controversial, technique. Remember, if you propose that a new methodology or technique is superior to a standard one, many who have been comfortable using the standard technique may find your methodology threatening to their own professional standing and reputation. The description of results is usually an easy section to write, although you might have to avoid the tendency among recent Ph.D.s to justify every finding via tedious logical arguments or overly long corroborative references. You are now writing for a specialized audience who will presume you have certain background knowledge common to all physical geographers. You need not, therefore, cite Newton every time you invoke gravity as a causal agent of mass movement. I write the abstract after the paper is finally complete, because in some cases I am not completely sure what to include in the final draft until it is finished.

If you are like me and many others, you will not complete a draft before "burnout" strikes, and you cease writing. I find it helpful to have several articles in the conceptual stage of creation at the same time, to avoid serious writer's block. If you find that you are simply tired of writing about a subject, you can then go off and revise a different paper in a later stage of creation, or perform data analysis on a research topic in the embryonic stage. Alternatively, look through your slides for the ones you wish to have made into prints for that fourth article you have in mind. No matter how interesting and scientifically sound a given topic may be, I doubt that you will have the energy and concentrated attention—not to mention the necessary span of time—to carry it through from beginning to end without interruption. So, avoid the frustration that results by having several articles in your files, on your desk, or in the computer at one time, in different stages of completion. This allows you to work on *something,* even if it is no more than the graphics or tables for an alternate paper, or conducting the library work associated with a future literature review section. Some sense of accomplishment is very important to writers in the early stages of their publishing efforts.

SHOULD YOU COLLABORATE ON RESEARCH?

I have assumed up until this point that you are working in isolation to produce a sole-authored paper. But how often is this the case, and how realistic an assumption is it? In reality, you probably will find yourself frequently collaborating with students, your past advisers, and faculty colleagues. Each of these situations can be worthwhile and is often necessary for work in physical geography that transcends subdisciplinary boundaries. Difficulties do arise, however, when you collaborate with different individuals. I will examine the major issues of collaboration as I have experienced them. Your situation may be very different, but I believe that these past efforts in collaboration touch upon many of the benefits and difficulties you are likely to encounter.

Collaboration with your doctoral or master's thesis adviser is usually the first such effort you will experience. In some ways it is not a true collaboration, in that you have not produced portions of the research jointly and then merged them. In essence, you have conducted the research yourself under the guidance of members of your doctoral committee. Your dissertation adviser, and any other committee member you feel has contributed intellectually to the creation of the paper under collaboration, may be listed as coauthors. On the basis of my own experience and those of friends and colleagues, I believe that it is typical, and usually traditional, for the doctoral student to be listed as the first author, unless the adviser created the conceptual framework and generated the data, which were then handed over to the student for analysis. A general rule of thumb would state, however, that any coauthored papers resulting from your dissertation will probably have you as lead author.

The question of authorship order for a paper resulting from a master's thesis is more equivocal. Frequently, master's theses are data analysis exercises in which the adviser provides the idea, the theoretical framework, and even perhaps the data set. In such cases, a paper resulting from a master's thesis might well have the faculty adviser as first author and the student second. For the student to achieve first authorship, he or she should have contributed some creative component to the research in question beyond what was initially suggested and/or outlined by the faculty adviser. For example, I have been a coauthor on ten papers resulting from master's theses for which I was the faculty adviser or a committee member. In the latter case, I was included as third author on five papers because of creative suggestions I provided in the conceptual stage of the thesis project; I also helped draft and review the paper during its development. In the cases of students for whom I was major adviser, four papers have resulted thus far. On three of these the student was first author, even though I wrote the papers and submitted them for publication (with their full knowledge and approval). In these papers, the major research results were an outgrowth of the students' efforts, above and beyond what was originally expected of them in their theses. On one paper, I was first author and the student second (again, with the student's knowledge, approval, and insistence). That paper resulted from an idea I had been developing for almost ten years, and on which I had written a research proposal during my own graduate school days. The student carried out the data collection under my guidance in the field, and analyzed the results as part of his research assistantship duties.[4] If another rule of thumb is needed here, let it be this: Don't let your ego get in the way; unless you genuinely believe that the research in question was yours all the way, and the student's role was that of lab assistant without a major creative component, assign first authorship to the student. The student worked long and hard to create the thesis and, in nine out of ten cases, deserves first authorship.

Your first real collaboration begins when you initiate or become part of a research program with colleagues. Collaboration in physical geography is frequently necessary because of the complexities of the natural environment we attempt to examine and describe. It is very difficult to examine geomorphic processes, for example, without at least considering the role of soils, vegetation, climate, and in some instances animals. Some of these variables may play a sufficiently minor role, so you can rule out their influence. It is not unusual, however, to find that at least one of these variables is complexly intertwined with the geomorphic processes under examination; you may find yourself as a geomorphologist needing the opinion and expertise of a biogeographer, a climatologist, or a soils geogra-

[4]In this case, we were collecting tree-ring data; see Butler et al. (1986) for details.

pher. Collaboration may also occur with colleagues who have expertise in certain techniques you do not, such as remote sensing, geographic information systems or an advanced quantitative technique.

To illustrate how collaboration can work, I will describe my own research with Dr. George Malanson, a biogeographer at the University of Iowa. George and I met when we were both hired in 1982 in the Geography Department at Oklahoma State University. As we learned about each other's background and research, it became apparent that we had a mutual interest in natural disturbances and their effect on vegetation. My interests focused on the interaction between mass movement processes and vegetation, and particularly on the reciprocal relationship between snow avalanches and vegetative patterns. George was interested in the effects of fire on vegetation, particularly in modeling successional patterns following brush fires of differing intensity and frequency. We had a mutual interest in the interaction between flooding and vegetation. I was interested in the effect of flooding on tree-ring patterns in woody vegetation, and George had examined the seed-dispersal aspects of flooding in his master's thesis. We each had a strong interest in fieldwork, but also brought different strengths to the collaboration: George possesses excellent quantitative and computer modeling skills, and I had wide experience in the application of absolute and relative dating techniques, particularly in tree-ring analysis, lichenometry, and palynology.

With these areas of mutual interest, we began discussing research topics that might incorporate both our backgrounds and capabilities. We showed each other slides and photos of our recent study areas, and tossed around ideas that might generate interest from funding agencies. In the course of "swapping slides," George noticed the snow-avalanche paths I had worked on for my master's thesis. The paths cut vertical swaths through mature coniferous forests in the Rocky Mountains and are colonized by a wide variety of deciduous shrubs and herbaceous plants. George commented that the avalanche paths look similar to artificial fire breaks and fuel breaks designed to prevent the spread of forest fires. The moment was magical and almost cartoon-like, as imaginary light bulbs illuminated over our heads at the same time: "Say, do you suppose that avalanche paths could be utilized as fuel breaks in the event of a forest fire? What influence does avalanche frequency have on the vegetation path, and how does that affect the path's ability to act as a fuel break? What effect do avalanche magnitude and snow distribution within the path have on that ability?" We had a topic of mutual interest, and we began to examine how each of our areas of expertise could help answer these questions. We wrote several research proposals to secure funding to examine the role of avalanche paths as fuel breaks in a forest fire modeling scenario, and received funding from the Association of American Geographers and the Burlington Northern Foundation to conduct a summer of fieldwork on the topic.

We carried out the fieldwork in Glacier National Park, Montana, where I had done the fieldwork for my master's thesis. We worked in close proximity in the field, but gathered separate data sources reflecting our individual capabilities. George and a field assistant collected data on species present and fuel load along transverse transects across avalanche paths, whereas another field assistant and I collected tree-ring and topographic site data along the transverse and longitudinal margins of the avalanche paths. We always met at lunchtime to discuss questions, problems encountered in sampling, suggestions for more efficient data collection, and so forth.

Upon returning from the field, George analyzed the species and fuel load data, and I processed and analyzed the tree-ring and geomorphic data. Whenever either of us saw interesting items emerge from the data analysis, we would get together to discuss the implications for our own and each other's analysis and results. In this process, ideas were exchanged freely and we never worried who suggested what or provided the insight that led to successful conclusions.

Following data analysis, we both wrote at the same time, with George describing the results of the fuel load and species richness data, while I wrote on the geomorphic implications and reconstruction of avalanche frequency and magnitude. When stumped on a section, we asked each other for advice and input and then returned to writing. Upon completion of a draft, we would swap papers and provide our own additional data, thoughts, and input. The final stage of paper preparation then included additional drafts incorporating each other's suggestions and criticisms.

We have never had any question about who should be first author on a specific paper. If the paper deals primarily with fuel loads, species richness, or patterns of vegetation on avalanche paths, George is first author. Such papers have been published in *Physical Geography, Journal of Environmental Management, Forest Ecology and Management,* and *Great Basin Naturalist*. Papers focusing on avalanche chronologies and magnitude, tree-ring analysis, and the type and geomorphic efficiency of avalanches in the study area have been published with me as first author in *Mountain Research and Development, Journal of Glaciology,* and *Professional Geographer*.

We mutually discuss and decide which journal to target for a specific paper; the first author submits the paper and follows it through the editorial and review process. If an editor calls for revision, the first author attempts to resolve and clarify the various issues raised in the reviews and then sends the paper to the other author. The second author examines the revisions of the first, adds his own suggestions and criticisms, and sends the paper back to the first. At that point, the second author tells the first to use suggestions as he sees fit and to resubmit the paper at his convenience. Using this methodology, we have been relatively successful in our publishing efforts, have never had a major disagreement about which revisions are necessary, and have never had problems over the order of authorship. Perhaps it is because we like and respect each other that we have been able to avoid any major problems in our collaboration; I think the key is to avoid doing to your collaborators what you would not wish them to do to you. The golden rule was never more appropriate than in determining how collaborative efforts succeed or fail. Our collaboration continues today, and we currently have several papers in various stages of development.

HOW DO YOU KEEP TRACK OF "PAPERS IN THE MILL"?

Once you have submitted a manuscript for which you are the responsible author, you must keep track of it, deal with the journal editors, and know when something has gone awry with the review process. To simplify these tasks, I advocate the use of "The Matrix" (Table 1), which allows you to keep track of a manuscript's progress toward publication, every step of the way. You can develop your own matrix to fit your needs, and you can keep it on a piece of paper or on a blackboard. I have the Matrix on a blackboard visible from anywhere in my office, so that at any moment I can consult it to determine exactly the status of any paper in question. Table 1 illustrates how I use the Matrix.

You should always keep a record of when you submitted a manuscript, and if and when it was acknowledged by the journal editor. *Keep copies of everything*: the paper, figures, and tables; the cover letter to the editor you included with the paper; and the editor's acknowledgment of receipt. (Some editors are more prompt to acknowledge receipt than others. If you are really worried, either call the editor directly, or mail your manuscript using certified mail, return receipt requested.) Do *not* be afraid to call editors. Their job is to process and rule on manuscripts. Therefore, they are dependent on you as a potential author for the health and viability of their journal. Call them if you have questions about the reviews the paper received, if you would like clarification of conflicting reviews, or if you would like greater insight into why revisions are necessary or why a paper was rejected. If revisions are requested, finish them as rapidly as possible and resubmit the paper. As Brunn (1987, 1988) suggests, many potentially publishable papers are never published simply because an editor asked for revisions and a resubmittal for a second round of reviews. Admittedly, the paper may not be accepted and published even after an additional round of reviews, but it surely will never be accepted as long as it languishes in your files or on your desk, so *resubmit*. If the paper is ultimately rejected, identify a second choice of journal outlet and submit again after trying to resolve questions and problems pointed out in the initial reviews.

Use the Matrix if a paper is resubmitted. It is flexible and can accommodate these contingencies. Use it after the paper is accepted to keep a record of when the

Table 1
A suggested matrix format for following "Papers in the Mill"

Paper's Working Title	Target Location	1[a]	2[b]	3[c]	4[d]	Date Submitted	Date Ackn.	Need Revision	Rewritten?	Resubmitted?	Accept?
										Revision Ackn?	Proofs?
Glacier Park Solifluction Lobes	*Zeitschrift für Geomorphologie*	GM	X	X	X	10–22–87					
Ram River Permafrost	*The Geographical Bulletin*	—	X	X	—	1–9–87	1–15	—			Yes / Yes
Glacial Hazards in Glacier Park	*Physical Geography*	—	X	X	—	2–16–88					
Colorado Tree Rings on Avalanche Paths	*Environmental Geology & WS*	JV	X	X	X	9–9–86	10–28	—			Yes
Slide Lake Landslide Dam, Glacier Park	*Mountain Research & Development*	GM	X								

[a] *Coauthor? If so, who?*
[b] *Graphics finished?*
[c] *Paper written?*
[d] *Paper revised and approved by coauthor?*

paper was submitted, and if you have received galley proofs. If, long after you have returned the galley proofs, the paper has not yet been published, call the editor again. A problem may have occurred in scheduling, your proofs may have been lost, or any number of other possible snarls may have developed; nevertheless, just knowing will give you peace of mind, so find out!

CONCLUSION

There is no formula for success in conducting research and writing papers for publication in physical geography, except for hard work and persistence. It may seem at times as though every editor and reviewer in the country is conducting a personal vendetta against your research efforts. Nonsense! You not only *can* do research, you *must* do research if you wish to succeed in the discipline. You *will* succeed in publishing the results of your paper if you don't give up, if you work hard, and if you are realistic in your expectations as to where a paper should be submitted. I wish you luck and success in your efforts.

REFERENCES

Brunn, S. D. 1987. Personal and general publishing policies of geographers. *Terra* 99(3):155–65.

———. 1988. The manuscript review process and advice to prospective authors. *Professional Geographer* 40(1):8–14.

Butler, D. R. 1986. Snow-avalanche hazards in Glacier National Park, Montana: Meteorologic and climatologic aspects. *Physical Geography* 7(1):72–87.

———. 1987. A Pinedale/Bull Lake Interglacial paleosol and its implications, central Lemhi Mountains, Idaho. *Physical Geography* 8(1):57–71.

———, Oelfke, J. G., and Oelfke, L. A. 1986. Historic rockfall avalanches, northeastern Glacier National Park, Montana, U.S.A. *Mountain Research and Development* 6(3):261–71.

Gaile, G. L. 1985. Rankling at rankings. *Professional Geographer* 37(1):62.

Kenzer, M. S. 1988. (Ex)cit(e)ation analysis. *Area* 20(1):73–74.

Lee, D. R. and Evans, A. S. 1984. American geographers' rankings of American geography journals. *Professional Geographer* 36(3)292–300.

———, and ———. 1985a. Reply to comments on "American geographers' ranking of American geography journals." *Professional Geographer* 37(1):62–63.

———, and ———. 1985b. Geographers' rankings of foreign geography and non-geography journals. *Professional Geographer* 37(4):396–402.

Peet, R. 1985. Evaluating the discipline's journals: A critique of Lee and Evans. *Professional Geographer* 37(1):59–62.

Turner, B. L., II. 1988. Whether to publish in geography journals. *Professional Geographer* 40(1):15–18.

On Writing and Publishing in Human Geography: Some Personal Reflections

L. S. Bourne

Skeptical readers may wonder why we need an article on writing and publication in human geography. Instructing a professional geographer, and particularly an academic, how to prepare and publish a scholarly paper is rather like telling someone how to breathe. We all do it, and we do it in our own very personal way and with our own unique style.

Yet there are indeed a number of common issues and concerns and some canons of good scholarly writing and publication that need to be voiced, primarily (but not exclusively) for the benefit of new entrants to the field. Some of these concerns must seem obvious. On the one hand, the quality of academic writing in human geography is widely perceived to be low, and the quantity is concentrated among relatively few of our academic colleagues. On the other hand, many young scholars in the field often view the process of publication in professional journals as an intimidating exercise set within a largely unknown and complex administrative labyrinth.

This chapter, as will be immediately apparent, is not itself the outcome of systematic and scholarly research. Nor is it intended as a model of clear writing or thinking. Rather, it is essentially a personal perspective, based on a retrospective review of a random sample of lessons learned from nearly twenty years of experience with the sticks and stones of academic criticism and the sometimes chaotic world of scholarly publishing. The paper is directed primarily at young scholars and practitioners and emphasizes the business, style, and conundrums of the publication process. It is not intended as a writer's manual, since there are many of these readily available. Nor does it review the finer points of journal formats and procedures, as these have been articulated clearly in recent papers by Brunn (1988) and Hanson (1988). Nor is it a review and assessment of the alternative styles, contexts, or approaches to research in human geography. Others have already provided such reviews in considerable detail (Johnston 1986; Eyles 1988).

It is frequently stated that scholarly research without formal publication is essentially a waste of time and social resources. The dissemination of ideas, of research findings, and of critical evaluations is an essential component in the process of advancing human knowledge and understanding. Publication also acts to create and nurture professional "communities of interest," and to display these interests to the population at large. It is also a necessary prerequisite for career advancement in the professional academic world, at least under contemporary institutional rules and regulations.

Typically, the dissemination of current research findings takes place in printed form through the professional journals published by individual academic disciplines, professional associations, nonprofit organizations, and private-sector corporations. Also, it should be noted that most journals are intended to be money-making operations for their publishers. This means, among other things, that the competition among journals for good articles has increased, which in turn has expanded the range of choice available in selecting an outlet for one's work. This chapter focuses on only one part of this complex and fascinating business of journal publication: the process of preparing, submitting, and resubmitting journal articles.

Despite grand pronouncements on the purpose of publication by learned societies, professional associations, and special interest groups—especially when they are seeking financial support from their respective granting agencies—the practice of journal publication tends to assume the nature of a combined game of chicken and chance. The criteria relating to academic appointments, tenure, and promotion clearly place the highest priority on the quantity of scholarly publications, either books or journal articles. Individuals are therefore encouraged to publish early and frequently. Given these imperatives, combined with the uncertainty of the peer review process and the absence of a clear consensus on what constitutes an acceptable standard for publication in terms of quality or quantity, many academics tend to play the game by attempting to maximize the rate and breadth of their publication record.

STARTING OUT

For many young scholars, the major hurdle, as DeSouza (1988) has recently noted, is getting started. Many desist from submitting a paper for publication because of a lack of confidence in the quality of their own work, or out of a fear of being rejected for reasons that were not anticipated, or out of a broader fear of the unknown. My first recommendation is to overlook these concerns. Get in the habit of writing and preparing articles for publication—but only submit those articles that are carefully crafted and based on solid analyses and thoughtful research. The publication process itself is valuable in that it forces us to keep in touch, to stay on top of the literature, and to retain a critical eye for quality research and writing. I have seen too many colleagues who are competent and have something worthwhile to say to the profession at large disappear from the academic literature precisely because of a failure to maintain momentum in terms of publication. Their places in the journals are often taken by others who have much less to contribute.

Most doctoral dissertations remain unpublished, which may on the whole be a good thing, although it does raise the broader question of whether the standard dissertation format represents a huge waste of human intellectual effort and collective social resources (Halstead 1988). Nevertheless, many dissertations contain the seeds of two or three respectable journal articles. Newly crowned graduates should attempt to prepare and submit these articles within twelve to eighteen months of completing their doctorates. For subsequent promotion and tenure, however, they must also move beyond the confines of the dissertation topic by extending the analysis or the conceptual arguments into new and different areas. At this point momentum is important, as are hard work and a sense of confidence, dedication, and enthusiasm to explore interesting intellectual questions.

At the same time, and given that we approach the task of writing for publication with widely differing backgrounds and abilities, we should not expect every article to unfold in the same way or in the same time interval. Sometimes a plausible article will appear from one or two writing sessions; at other times it will evolve only after a long struggle. In still other instances, the entire effort may lead to a dead end. Weaving a complex set of ideas, premises, and analyses into a cohesive written product is never an easy task.

In all cases, I recommend that new authors "sit" on the completed article for a time, and then approach a second draft with what should be a clearer and somewhat more detached perspective. In some instances—for example, when faced with an immediate deadline—the process may be sharply truncated, but it should still be carried out. It is also of immense help to circulate papers, among colleagues whose opinions you value, for their critical assessment before submission to a journal. If colleagues at your own institution are not helpful, try reaching others through BITNET, NETNORTH, or some other form of electronic mail network. These are often the most useful comments and criticisms that you will receive. Finally, the most difficult part of a paper to write is often the introduction. If and when you face this barrier, leave it until the end. You can then adjust the content and style accordingly.

CRAFTING THE ARTICLE

It should now be clear that there is no single path to the preparation of a good journal article, nor is there a single model of how an article should be written and organized. Each author is different, the journals differ in approach and expectations, and academic backgrounds and writing abilities vary widely. Nor do the social sciences in general lend themselves to uniform packaging in terms of the format, styles, or specifications of journal articles. Having said that, however, it is important that young scholars be aware of the variety of approaches and strategies that are commonly followed.

The task of designing an article for publication is essentially an individualized exercise. Most papers begin with the germ of an idea, an interesting question, a policy problem or premise of theoretical concern, or a challenge to investigate a new avenue of empirical research. In some instances the idea is suggested by an external source, such as through an invitation to a conference or to accept a research contract.

That idea in turn poses a set of hypotheses or assertions to be evaluated and calls forth a suitably revised theoretical framework. Once this context is in place, the eager researcher constructs an analytical design, outlines a set of arguments, and selects (when appropriate) a body of empirical data on which to test the ideas, hypotheses, and theories defined at the outset. Alternatively, one can begin with an empirical analysis of a set of observations and then undertake to construct a paper around the results of this analysis. The former approach starts with the context and fills in the body of the paper in subsequent stages, while the latter begins with the content or analysis and proceeds to construct a suitable conceptual box or framework around that analysis.

Where one starts the process is less important than what goes into each component, and how those components fit together to form a logical and synthetic whole. Much of the empirically based literature in human geography conforms to a research design that incorporates both an inner and an outer circuit, in terms of the basic research strategy, as suggested in Figure 1. The inner circuit focuses on the empirical analysis itself and specifically on the research problem and the hypotheses that one has set out to investigate. The outer circuit frames the specific analysis, providing the broad context that is the environment of ideas from which the analytical questions are drawn and to which the results of the analysis are to be generalized.

One does not have to be well versed in the philosophy of science or in the current methodological debates in the social sciences to detect the two key difficulties here. The first problem, as in any branch of scholarly research, involves the translation of ideas from the outer circuit, or conceptual context, into propositions or hypotheses that are analytically testable within the inner circuit. The second

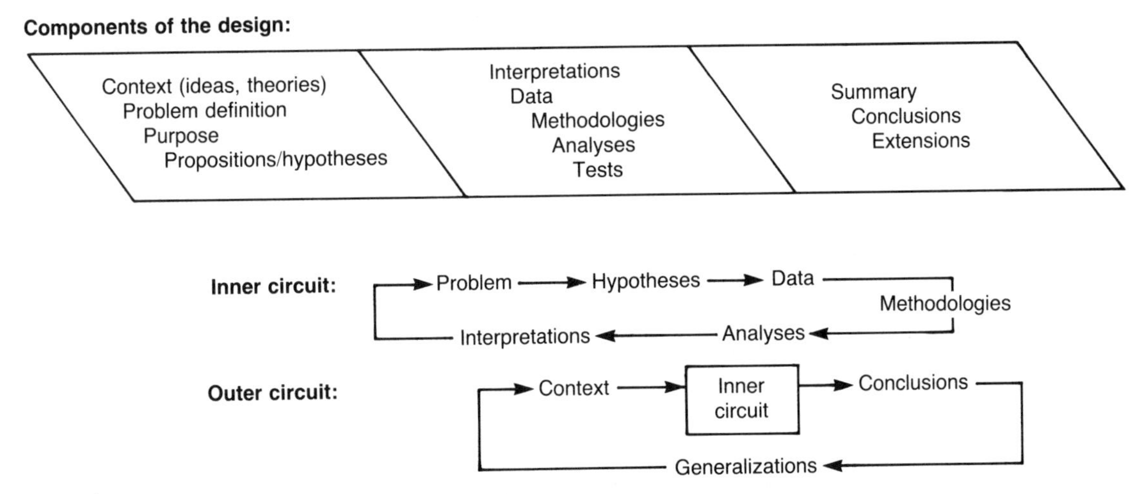

Figure 1
A typical research design in human geography.

involves the subsequent abstraction from the inner circuit, based on an evaluation of what are invariably limited empirical analyses, to the higher level of generalization necessary to address the issues raised in the contextual discussion. It is often in these translation phases, linking the general to the particular, and then the particular to the general, that many academic articles come apart. There is no simple solution to this problem, although it does help if, when writing an article for publication, one continually asks the question, "Do these analyses and these empirical results address the general issues raised in the introductory statement of purpose?"

SOME PRAGMATIC LESSONS

Moving from generalities to specifics, the first and perhaps the most important lesson is to write clearly and simply. Do not use three words or three syllables when one or two will do equally well. Avoid excessive jargon, technical language, and mathematics; use only what is necessary and sufficient to ensure that the average reader of that journal will understand your purpose and direction. Obviously there is a need to include certain key words, citations, or equations—call them a code language—for purposes of convenience and efficiency. Also, it is useful to let your particular peer group know that you know the relevant literature and that you are also a "paid up" member of that group.

Remember, however, that the intended audience is (or at least should be) much broader than your own immediate peer group. Ask yourself whether you are preaching only to the converted. In my view, you should try to strike a balance in the choice of vocabulary, writing style, and citations between the need for peer group recognition and the benefits of addressing a wider audience. For mathematical and data-rich papers, detailed derivations, models, and calibrations can often be assigned conveniently to an appendix.

The most common problem in academic publishing, however, must be the lack of clarity in the organization and flow of ideas within the body of the paper. As suggested in Figure 1, this is both a problem of logic and one of consistency in articulating the conceptual-analytical links involved in the research strategy. Editors frequently complain about the disjointed nature of the papers submitted to their journals, and particularly of the weak correspondence between the initial conceptual framework, the data and methodologies used, and the interpretations and generalizations that flow from the analysis (Hanson 1988). Most rejected articles are turned down for precisely this reason. This discordance may be the result of several factors: premature submission, a failure of self-criticism, deliberate distortion, or simply careless writing and sloppy thinking. In any case, authors have both a personal and professional obligation to ensure that the paper actually does deliver what it proposes to deliver.

WHERE TO PUBLISH?: SELECTING THE JOURNAL

Given the proliferation of academic journals in human geography and in those transdisciplinary fields of research in which geographers have a prominent role, selecting the appropriate journal as an outlet is nowhere near as straightforward today as it used to be. Each journal is different. As an outlet for the results of one's research efforts, the choice of a journal obviously depends initially on the nature and subject areas of the research, the focus of the journal, and the audience that one wishes to address. Although journals collectively reflect the kinds of research that we do, their policies and practices in turn steer research priorities within the discipline.

In the present academic environment the best strategy undoubtedly is to spread one's papers across a range of journals. We now have journals for those who are numerically inclined, for those who profess to take a political economy approach, for those who prefer applied research, and for those who specialize in subfields within subfields. New authors should spend time in the library to become familiar with the full range of journals available and their respective styles and contents.

Assessing the quality of these journals is not an easy task, particularly in such a diverse field of scholarly inquiry as human geography. There is no annual consumer report on journals. Yet there is an implicit ranking of journals within the discipline, based on a combination of the size of the audience (that is, the print-run), the latent demand for publication (the backlog or wait time), the awareness level of the readership (the citation indices), the reputations of the editors and previous contributors, the rejection rate (if known), and the apparent thoroughness of the review process. Authors all have their own quality rankings, although younger scholars may have to seek the advice of more senior colleagues to obtain "insider" rankings.

It goes without saying that the author should select a journal that best matches individual interests with the intended audience. Even so, the choice often involves a trade-off between the ease and speed of publication and the quality of the journal. Most academics know of journals that have little or no backlog and little or no serious peer review. The turnaround time from initial submission to publication (which varies from nine months to almost two years for the most prestigious journals) may be much shorter in these journals. Yet it is clearly advisable for new faculty members, especially those facing a tenure and promotion hearing down the road, to direct one or more of their early papers to a mainstream or "flagship" journal within the discipline, or to high quality journals in closely related fields. Tenure and promotion committees also have lists of such journals in each field. These committees increasingly discount the value of publications in peripheral or second-rate journals.

Some institutions indeed grant tenure on the basis of a faculty member having some minimum number of papers (say, five) accepted for publication by any of the top-ranked journals in that discipline. This of course puts additional pressure on the candidate to target the submission of articles, as well as on the overworked editors and reviewers of the journals involved. An obvious danger in this approach is that the core of the evaluation process in tenure reviews will be shifted inadvertently to the detached and sometimes less informed external reviewers in the peer review process. In any case, it is usually wise to select a high quality journal over a low quality one, even if it slows down the rate and volume of publication. And, of course, an initial rejection does not preclude resubmission elsewhere.

As Turner (1988) has suggested, publication in a non-geography journal often brings greater recognition and a wider audience than publication in a mainstream geography journal. This, unfortunately, is also characteristic of many of the smaller social science disciplines, but perhaps is more symptomatic of the diffuse nature of the geographical community. If carried too far, this tendency would lead to still further fragmentation of our already subdivided discipline.

The solution for all geographers is clearly one of balance, selecting the best and most relevant journals as outlets both within and outside the discipline. New scholars, however, should avoid the more esoteric and obscure journals, at least until they are better known and preferably tenured.

READING YOUR AUDIENCE

Aside from tenure and promotion considerations, the impact and usefulness of an article depend in considerable part on reaching the appropriate audience. As the number and frequency of journals have increased—and they have increased far more rapidly than the size of the academic community and its ability to consume information—audiences on average have tended to become smaller, less well defined, certainly more fragmented, and probably more specialized.

Consider journal editors and external reviewers as the first audience. Most editors are known, and in any case they represent a captive audience. All editors maintain extensive lists of reviewers, typically classified by subject area, location, reliability, past reviewing experience, and type of specialization. Editors may also scan the author's bibliography to find suitable candidates for reviewers. Despite the potential length of these lists, only a small pool of reviewers is qualified to review a paper on a relatively specific theme in a subject as thinly spread as geography. This allows the author a reasonable guess at who the reviewers might be. It also increases the probability of facing an inappropriate or unsympathetic reviewer, and you should not be surprised if that happens. It is almost certainly a waste of time to try to anticipate precisely who might review your article for either journal A or journal B. Nevertheless, the journal's recent record of publications and reviews should provide some clues as to the kind of reviewer likely to be selected. It does not harm one's chances to make minor adjustments in the paper in anticipation of the possible reactions by these reviewers. At least one should cite the right people in the list of references, those who have published significant and relevant work in your field, to pay homage to one's peers.

Of more importance in the longer term is the readership of the journal. Unfortunately, this is perhaps the most uncertain element, since we really know very little about the readership of professional journals in human geography. We do know, however, that the audiences differ substantially in size, interests, subject area, and qualifications, depending on the focus, country of origin, and quality standards of the journal. These audiences are nonetheless of central importance, both as consumers of your product and as the ultimate peer group. Write your paper as if it were addressed to them personally.

Invariably, journal articles are returned to the author with suggestions for—at a minimum—minor revisions and clarifications. This is to be expected, and on the whole it should be a positive learning experience. Remember that most reviewers will at least want to be seen to be providing a thorough and critical evaluation. Finding negative things to say about a paper, therefore, is considered to be part of that job, even if they have not read the paper thoroughly. More positively, the practice of constructive criticism and informed debate is an essential component in the canons of scholarly publication and intellectual discourse in any field of scholarly enquiry.

It is advisable to respond to these criticisms and comments as if they were all substantive and serious. You should not despair, swear, or respond with an inflexible "I will not budge an inch" attitude, even if you feel that the reviewers' comments have been unfair or unrepresentative. Instead, respond positively, making as many changes as you feel are possible without compromising your approach or integrity.

Do not enter lightly into disagreements with editors. If you must disagree with them, make your points clearly and calmly; do not engage in lengthy and acidic correspondence that benefits no one. Editors do expect that revisions will be more than cosmetic and will tilt the paper however slightly in the direction that the reviewers have suggested. Then resubmit the paper, if necessary with a note of explanation on any suggested revisions that could not for one reason or another be incorporated. In those instances in which you disagree with the reviewers' comments, consider this as an involuntary opportunity to clarify further your position and the line of argument within the body of the paper. Decide how far you will go in revision, and then stick to that position. Frequently, however, the comments of external reviewers tend to cancel each other out, leaving the editor in limbo and the author no further ahead in preparing revisions. In this instance, one should at least acknowledge in the revised paper the diversity of views on the issues under discussion.

The specific and most appropriate responses to criticism through peer review obviously depend on the nature of the critical comments. Some reviewers will focus on data and methodology, others on the conceptual framework or theoretical apparatus, and still others on the internal consistency of the arguments and interpretations. Almost all referees will ask you to do more work: more data, more analyses, and additional references. Again, this is to be expected, and one should respond as much as is possible within the constraints of the research on which the paper is based. Looking on the positive side, these suggestions for additional work may form the basis of a subsequent paper or research project.

New scholars also face a challenge simply because they are new to the field and thus are relatively unknown in the academic community at large. This has both disadvantages and advantages. One obvious disadvantage is that senior or well-known scholars sometimes benefit from an easier ride through the peer review process, in part because of their established track record in publication, and in part because of their position in the network of experts on a specific subject. Precisely because they are well-established names, senior scholars may also be able to get away with fewer citations to the existing literature and more to their own work.

Recent Ph.D.s, on the other hand, must earn their place in the sun, one paper at a time. They obviously cannot rely on past performance (Kenzer 1988). On the other hand, in some circumstances new scholars may receive the benefit of the doubt to compensate for the lack of a track record in publication, and they are likely to have fewer opponents gunning for them by name alone.

In dealing with journals, remember that most editors are not obstructionists. All editors want solid, first-class articles for their journals. Assuming they receive a steady flow of articles, on which their journal's survival ultimately depends, most editors are primarily concerned with maintaining high standards rather than selling a specific basket of academic goods. It has been my personal experience in geography that most journal editors are extremely sympathetic and helpful. They have a difficult and often thankless job, and may only become household names themselves when serious conflicts, complaints, or disputes arise.

Reviewers are a different matter. Most, it should be emphasized, do their difficult task admirably. But they vary widely in personality and competence. Some are delinquent in responding to requests from editors for comments, causing unwarranted delays and undue frustrations. Some are picky; some are mean. Others simply fail to put in the effort necessary to complete a thorough and thoughtful review. Still others use the peer review process for the inappropriate purpose of polishing their own pet research paradigm or for scoring points in some wider debate within the discipline that may not have anything to do with your paper. Some were simply bad matches for the article in question. All one can do is grin and bear the process.

THE REJECTION LETTER

The most severe form of criticism, the rejection of a paper, is of course more difficult to accept. On reflection, however, the author may in fact agree with the reviewers' criticisms and recommendations, and the editor's decision. If so, an important lesson should have been learned. If the author disagrees, it is usually not worth the effort to register a protest. The options then are threefold: (1) revise substantially and resubmit; (2) revise and submit elsewhere; or (3) file the paper away. Which course of action is most appropriate in any particular situation depends on the quality of the initial paper, the nature of the rejection notice, and the importance the author attaches to publication on that subject in that journal at that time. Some of the best articles in the discipline were initially rejected.

Two further points should be kept in mind. First, it is inadvisable to submit exactly the same paper to another journal, in the hope that another field may be more fertile and another set of editors and reviewers more accommodating. There is some overlap of reviewers, and there is an informal editors' network in publishing. Nothing annoys a referee or editor more than an attempted end-run around a rejection notice. This of course excludes those situations in which papers were rejected because the editors considered the subject matter of the article to be more appropriate for another journal.

Second, remember that you are not alone. Everyone in your field, no matter how senior or distinguished, has received a rejection letter of one type or another at some time in their career. It is not the end of the world or of your career. Extract what positive lessons you can from the experience, and move on. Also remember that the probability of receiving a rejection notice is significantly higher in fields of research that are relatively new and untested. These are precisely the fields that tend to attract young scholars.

COLLABORATIVE RESEARCH AND COAUTHORED PUBLICATION

It has become increasingly common for human geographers to participate in team research (often, but not necessarily, multidisciplinary research), and to publish collaborative papers based on that research. In my view, and for the most part, this trend has been a positive one. It encourages a wider exchange of ideas and practices on the one hand and allows for the accommodation of differing research styles and approaches on the other.

The collaborative publication of journal articles does pose a number of practical questions that cannot be ignored. The major concern is that joint authorship makes it difficult if not impossible to separate the relative contributions of the respective authors. At the time of tenure and promotion decisions, this consideration is not insignificant. Tenure committees tend to discount coauthored papers, roughly in proportion to the number of authors involved, and particularly if the senior author is the past or present doctoral supervisor for the other author (or authors). Young scholars are therefore strongly encouraged to publish at least some of their collaborative research on their own, of course with the consent of their associates and with full recognition of their contributions to the joint research project.

Coauthored papers also pose a minor but nevertheless tricky question of protocol: whose name should appear first? Aside from ego considerations, this is important only to the extent that citation indices often use only the first name in a multiple-author paper. There are four principal ways to decide on the ordering of names: (1) in terms of the relative size of each contribution or of the initiator of the research; (2) by alphabetical order; (3) by seniority; or (4) by a flip of the coin. To reduce the possibility of disagreements, a decision on the ordering of names preferably should be made before the paper is written or early in its preparation. If the relative contributions of the authors change as the article develops, a reordering of the names may be appropriate. If not told otherwise, most readers will assume that a nonalphabetical list means that the first author initiated the research and/or made the largest contribution.

THE PUBLICATION MACHINE: THE ARTICLE KIT

Some of our more cynical colleagues would argue that the increasing pressure to publish as frequently and as widely as possible, combined with the standardization of journal formats (and widespread word-processing capability), has produced extensive and repetitive publications, and perhaps a lowest common denominator in the design and content of academic articles. One result has been the appearance of what might be called an "article kit"—a prepackaged and homogenized format for the rapid production of one journal article after another. I am certainly not suggesting that our younger colleagues adopt this model, but a brief example may help to explain how and why some human geographers publish so frequently.

One such format might take the following form. The paper begins with an introductory paragraph setting the context and identifying the problem. The problem is frequently defined through a pointed criticism of someone else's work—what might be called the "straw person principle." That criticism is typically supported by a carefully chosen quotation from another "big name in the field," or perhaps from someone outside the discipline altogether, or—even better—from an obscure source no one else has heard of. This establishes what we might call the "indisputable reference principle." Next we might expect a brief review of the literature, selectively chosen to support the case made by the author or to establish a counter-argument that is to be disproved. At this point, or perhaps in the next section of the paper, one would expect to find a short list of hypotheses, defined on the basis of the indisputable reference and aimed at the straw person.

It is likely that the paper would then launch into a description of the data and the methodologies to be used. Ideally, it also should be possible for the author to claim that no one has previously used these data—at least not efficiently—or that the methodology in question has never before been applied to this problem, or this data set, or this particular locality. This might be called the "unique applications principle."

Next, the author would review the analytical results and draw out the inferences and generalizations that support the hypotheses outlined earlier. If the straw person principle is the target, then the counter-argument or antithesis is inevitably dismissed with firm dispatch. The last section of the kit then draws the conclusions together as they relate to the specific analytical framework and the broader context established initially. Often, however, these generalizations vastly exceed the scope and utility of the empirical base, if not of the overarching conceptual framework. Once you have developed the style, mastered the technique, and collected a file of references, the papers then begin to roll off the word processor as though from an assembly line.

The preceding scenario is not intended to cast doubt over either the process or the standard format for publication of academic articles. Rather, it is intended to alert the new scholar to the weaknesses of the review process, the inherent contradictions in current academic evaluations, and the dangers contingent on the rule "publish often and as often as possible."

Too frequently, quality is sacrificed on the altar of numerical supremacy. This is unfortunate not only for individuals but for the discipline as a whole. For individuals it tends to produce the false impression that they have accomplished a lot. It erodes others' impression of the quality of your research. It also tends to erode your own sense of quality, of what is a worthwhile scholarly contribution, and this attitude is then passed on to the next generation of students. Lower quality papers become not just possible, and publishable, but a self-perpetuating common standard. The discipline suffers accordingly.

It seems to me to be worth repeating some of the positive lessons identified in the preceding discussion.

1. Never submit a shoddy piece of work. Papers should be well written, carefully edited, logically organized, and internally consistent. Do your homework first. Do not rely on reviewers and editors to edit and rewrite your paper. Let the reviewers concentrate on the contents, and thus on the merits of the paper itself. Learn to practice critical self-editing.
2. Conform to the journal's published style guidelines. Try to prepare a paper—at least the final draft—with a single journal in mind. Otherwise you are asking for further headaches, longer delays, unnecessary revisions, and possible outright rejection.
3. Do not overburden the reader with superfluous detail and unnecessary repetition. When in doubt, leave it out. A short good paper is better than a long good paper; and this is even more true if the paper is not good.
4. Unless you are writing a bibliographic essay, do not reprint your reference card file (or diskette) in the paper. And do not cite references that you have not read—they could be your undoing.
5. Do not oversell your results; be accurate, balanced, and most of all honest in assessing the merits of your own work.
6. At the same time, do not overemphasize the shortcomings of the paper or the analysis.
7. Ensure that the building blocks of the paper actually do link directly to each other. For those students who have difficulty in maintaining a flow of ideas and a structured argument from beginning to end, I often suggest the use of a detailed, section-by-section outline. At times a traditional flow chart is useful, although this is clearly more suitable for some styles of research than for others.
8. Do not publish what is essentially the same paper two or three times in different journals, although parts of the supporting documents may be reworked for inclusion in different articles.
9. Maintain a healthy respect for the process of peer review, but do not allow it to intimidate you.
10. Remember that, ultimately, the quality of publication cannot be separated from the quality of research and of graduate training in the discipline.

Perhaps one of the basic problems facing the discipline, at least in human geography, is that there really are very few first-class articles published. Most are good, but too many have but a single novel methodological or theoretical twist or one good idea, which is then stretched into a feature-length article through the use of extensive padding and verbal puff.

The message is clear: Do not publish prematurely, and do not publish just for the sake of publication. A real dud of a paper will stick to you like the proverbial fly at a summer picnic. Our discipline has too many marginal papers, so try not to add to that accumulation. As our collective ability to consume published work has at best remained constant, while the volume of papers has risen sharply, the value of the printed word has declined significantly. We should not simultaneously depreciate our intellectual capital by accelerating that rate of decline.

REFERENCES

Blalock, H. M. 1986. *Basic dilemmas in the social sciences*. Newburg Park, CA: Sage.

Brunn, S. 1988. The manuscript review process and advice to prospective authors. *Professional Geographer* 40(1):8–14.

DeSouza A. 1988. Writing matters. *Professional Geographer* 40(1):1–3.

Eyles, J., ed. 1988. *Research in human geography*. Oxford: Blackwell.

Halstead, B. 1988. The thesis that won't go away. *Nature* 331 (11 February):437–38.

Hanson, S. 1988. Soaring. *Professional Geographer* 40(1):4–7.

Johnston, R. J. 1986. *On human geography*. Oxford: Blackwell.

Kenzer, M. S., 1988. (Ex)cit(e)ation analysis. *Area* 20(1):73–74.

Turner, B. L., II. 1988. Whether to publish in geography journals. *Professional Geographer* 40(1):15–18.

Work of Prostitutes or Missionaries: Producing Popular Regional Geographies

John A. Alwin

Ask nongeographer friends if they've read any good geography articles or books lately, and their only response is likely to be a blank stare. At best they may remember skimming through an article in a dated and dog-eared copy of *National Geographic* in the waiting room during their last visit to the doctor or dentist. Walk into any B. Dalton or Waldenbooks bookstore, or even many university bookstores, and try to find a book written by a geographer that is *not* a textbook. With few exceptions they will be absent from shelves and available only by special order.

If our field were physics or chemistry, and we dealt with wave-particle dualism or nuclear magnetic resonance, this situation might make sense. But geography deals with a wide range of subjects that interest the general public: people, places, cultures, landscape, the environment, and the past, to mention just a few. What we could share with the public is not only of great interest, but also timely and important.

There is a little bit of the geographer in everyone, whether people recognize it or not. North Americans have an interest in places both near and far. They wonder how areas came to be the way they are today, what distinguishes each from another, and how they are interrelated. Despite this interest in geography among the populace, professional geographers have made little effort to share their wealth of geographic knowledge and enthusiasm for the subject directly with the lay public.[1] Many in our field think of geography for the general public as unworthy of our efforts, a kind of geographic prostitution. And with a hyperspecializing discipline in pell-mell pursuit of more theoretical and quantitative approaches and enamored of new geographic techniques, only special individual and disciplinary efforts can prevent the public from being left behind in the geographic dust.

[1]For a rare and eloquent appeal for popular geography, see Peirce Lewis's 1985 presidential address presented at the annual meeting of the Association of American Geographers (AAG). Two years later, Ronald F. Abler offered a prescription in his presidential address for a healthier American geography (1987). In "What Shall We Say? To Whom Shall We Speak?" he called for increased dialogue among geographers in academe and between academic geographers and their students, teachers of geography in public schools, and colleagues in practice. No mention was made of a role for direct communication with the general public—an important and potentially huge audience.

Always-popular *National Geographic,* with more than ten million subscribers, is perhaps the best indicator of the public's fascination with things geographic. But lately publishers and nongeographer authors have sensed an unsatisfied appetite for geography. This sizable readership is illustrated by recent national bestsellers, including Rand McNally's "Places Rated" series of books, journalist Joel Garreau's *The Nine Nations of North America* (1981), former English professor William Least Heat Moon's *Blue Highways* (1982), and political economists Michael Kidron's and Ronald Segal's *The New State of the World Atlas* (1987).

On a local scale, volumes by nongeographers in a flurry of successful state geographic book series point to a home-focused topophilia.[2] Clearly, if geographers don't produce suitable geographic works for the general public, others will try—often under the banner of geography. This geography-without-geographers trend may impact adversely the discipline. In a worst case scenario, counterfeit geographies will be put forth as the real thing, adding to the already muddled public image of the discipline.

Scholars in such fields as history, economics, anthropology, and psychology write national bestsellers; why not geographers? And contributions need not be limited to books making the *New York Times* bestseller list. Other books, atlases, magazine articles, newspaper columns, radio and television programs, and films and videos all help to satisfy the public's interest in geography.

PROFESSIONAL GEOGRAPHERS AND POPULIST GEOGRAPHY

As graduate students, we are taught the rigors of academic research: intense and highly focused, and conducted by a dispassionate and unprejudiced researcher who strives to maintain a scientific detachment from what is studied. Emphasis is on data gathering, interpretation, and drawing of conclusions. Explanation—not description—is the primary goal.

Just as most advanced degrees say nothing about an individual's teaching abilities, they also say little about writing skills of any kind. The Ph.D. in particular is predominantly a research degree, often awarded to those with marginal writing abilities at best. After all, the emphasis is on what is said, not how it is said. The same just-get-by-with-it writing transfers well to many scholarly endeavors where, again, readers are primarily interested in the message, and the collective tolerance for minimal writing skill is great.

Not surprisingly, scholarly writing by geographers, as well as those in other disciplines, tends to be less than enthralling and is usually unacceptable for public consumption. Often it is mechanical, lifeless, and has as much personality as a government manual on income tax preparation. Chances are it is tedious reading even for a colleague with the same research interests.

Even ignoring the fact that our discipline holds populist geography in such low esteem, it is easy to understand why few professional geographers have ventured far from the traditional and familiar and into the different realm of researching and

[2]State geographic book series began with the Alaska Geographic Series, now in its fifteenth year. Additional state geographic series by other publishers have followed and now cover California, Oregon, Washington, Idaho, Montana, Wyoming, Utah, Colorado, Minnesota, Vermont, and New York.

writing geography for the general public. To do so may require a different approach to research and definitely demands an entirely new style of writing. If this were not so, the *Annals, Canadian Geographer,* and *Economic Geography* would be sold in magazine racks and would grace nightstands across the continent, and fans would line up at bookstore autograph parties to have us sign their copies of our latest scholarly tome.

Writing geography for the general public definitely is not for every professional geographer, but the more adventuresome from a wide range of specializations could join in and adopt it as an additional outlet for geographic expression. Timely topics and issues, such as immigration, minorities, the homeless, crime, the farm crisis, restructuring economies, changing lifestyles, politics, religion, environmental quality, international trade, transportation, and retirement, all have a geographic perspective and would be well suited to popular geography. Our profession has knowledgeable and perceptive individuals with a flair for writing. It is inexcusable that their work should be well known only to other geographers and, in a few cases, to those in other disciplines, but rarely to the public at large.

Regional geographies may offer our profession its greatest opportunities in the public arena.[3] If the general public expects anything from those in our field, it is that we tell them about places. That is what *National Geographic* does so well in its own way, and my guess is that most readers assume articles in that publication are authored by professional geographers. In reality, we may be asked to comment on regional pieces, but with few exceptions they are the work of journalists. Often these writers approach a place about which they have little prior knowledge, learn what they can in a limited time, write their piece, and move on to another corner of the world to begin the process anew. Their journalistic work and accompanying photography have earned the magazine its excellent reputation. Imagine what skilled geographers with intimate and long-term knowledge of places might do in regionally focused, popular magazine articles or in book-length regional geographies for lay readers—if geographers were sufficiently motivated to try.

When academic geographers do write nontextbook, regional geographies, these books tend to be published as scholarly works by university presses. Such titles are listed in the "New Scholarly Books" section of the *Chronicle of Higher Education,* qualify for reviews in professional journals, often are highly regarded by peers, and are considered pluses in tenure and promotion review. They may constitute important contributions to the discipline while enjoying some popular readership. But with their small press runs and unavailability at many bookstores, even within the corresponding study region, the impact of such works on the general public is minimal.[4]

This style of regional geography is substantially different from the unabashedly populist variety that is neither published by a prestigious university press nor

[3]For the best in repeated attempts by geographers to argue the case for regional geography, see Hart 1982. Hart's presidential address has special relevance to popular regional geography.

[4]Because of reduced institutional support and pressure to produce a profit, some college and university presses are turning to scholarly books with a heightened popular appeal as one means of expanding markets and boosting sales. Scholarly trade book hybrids usually cover general-interest subjects, commonly regional historical geography, and are designed and marketed differently from traditional scholarly volumes. This trend offers geographers some exciting opportunities. An excellent recent example of this variety of geography book is Borchert 1987.

announced in the *Chronicle,* is probably passed over for journal reviews, at best may be a "nonfactor" in tenure and promotion review, and may even draw snickers from fellow geographers. Yet a broad spectrum of area residents eagerly embraces popular regional geographies, which potentially can present geography to tens of thousands of people while simultaneously serving as a public relations vehicle for our discipline.

RESEARCH WITH A DIFFERENCE

Geographers conducting scholarly research are guided by their knowledge of our discipline and its perspectives and methods. By design, research is customized to the subject investigated, not to the audience for an eventual report on findings. Those researching geographies for the general public also are guided by their "geographerness," but the best popular geographies also are guided at the research stage by the interests of the intended audience. Readers of popular regional geographies primarily are those who call the area home.

It is essential that the researcher have some familiarity with residents. There is little hope of addressing reader interests if they are unknown. If residents don't sense the volume is especially for them, it will sit unsold even on the sale tables of area bookstores. Equally important, it is impossible to know a region well without knowing its people.

Geographers are well advised to select regions with which they already have a close relationship. It is not necessary that the association be lifelong, but the longer and more intimate the connection, the better the potential for good regional geography. Without this, the odds favor production of hollow "nonbooks" without heart or soul, reminiscent of encyclopedic inventories from an earlier era.

Assuming adequate familiarity with a place and its people, a starting point for popular regional geography is identification of two or three main themes that are key to the distinctive regional identity. A single theme is restrictive, invites the detail of an academic approach, and may produce a study that is more topical than regional. Trying to cover everything is likely to result in a superficial gazetteer-like regional geography. Once appropriate themes have been identified, settling on a preliminary organizational outline is necessary before research begins. At this stage it is not too early to begin visualizing the finished, printed, and bound volume.

Researching popular regional geographies has more to do with breadth than depth, and is characterized by both original research and synthesis of existing studies. Original research is an important component of popular regional geography, but the researcher must also rely on published works of others. Chances are there will be few sources by geographers. Instead, the researcher must become familiar with the regionally focused efforts of geologists, archaeologists, historians, biologists and botanists, soil scientists, agronomists, economists, political scientists, urban planners and others. It is no small task.

The researcher must also be prepared to consider a wide range of additional potential sources that normally are not referenced in academic studies. Industry and trade publications, radio and television programs, newspaper articles, regional magazine pieces, and even travel-related publications can be helpful. If the accuracy and objectivity of such sources are suspect, they still can provide insight and point the way to more reliable sources.

No matter how familiar you are with a region and its residents, you always can learn more, and there is no substitute for extensive fieldwork of the get-off-the-freeway-and-explore variety. Even with my smallest (50,000-square-mile) study, fieldwork included more than 10,000 miles of crisscrossing the region via most of its paved highways as well as many of its more challenging "local inquiry advised" roads.

For practical reasons, there is a tendency for academic geographers to use summers for fieldwork and the remainder of the year for indoor research and writing. But you can't truly know a region unless your familiarity with it extends to all seasons. Sights, sounds, smells, and the very pulse of a place change with the time of year. A summer season, tourist's view will not suffice.

Curiosity, enthusiasm, and enjoyment are keys to effective fieldwork and an inviting popular regional geography. Lighten up, loosen up, and leave your cap and gown back on campus. Most of the people you will encounter in the field are not interested in academic credentials or titles.

For me, interviewing people—a dying art among geographers—is the most exhilarating and personally rewarding aspect of research. "Locals" are the real experts on their area and the best possible source of information. For each of my regional geographies I interviewed scores of local residents, from farmers and migrants to rangers, mayors, mill workers, cowboys, and Indians. It is rare to find people who are not delighted to talk about themselves, their livelihood, and their region.

Residents' contributions to my studies are incalculable. Octogenarian Via Cooper of Jordan, Montana, who homesteaded on the eastern Montana frontier as a young bride in the 1910s, gave me insight into the settlement frontier that couldn't have been conveyed in the secondhand accounts of others. I learned more about grain farming, machinery, erosion, and marketing in the Palouse region by spending a day harvesting with the Mader brothers outside Pullman than I could have if I had spent a week researching those topics in the library at nearby Washington State University. After a day with the Amish family of Steve and Linda Kauffman in northwest Montana, I had an understanding of that religion and lifestyle that I could not possibly have gleaned from books and articles.

Be prepared, at least in small towns and rural areas, to be invited to stay for lunch or dinner. Numerous times I have had interviewees take out their albums and offer vintage family photos for use in a book. In some cases, initial contacts have led to long-term friendships. Such occurrences are a reminder that producing regional geographies for the public is a very personal endeavor that creates a bond between you and your region's residents.

I hesitate to use the word *interview,* since I doubt that I ever interview anyone. *Talk to* is what I think I do. People are put off by interviews, in which they are expected to respond to questions from a researcher. I've found it more productive and enjoyable for both parties if we carry on an informal conversation, often over coffee, the way one would with a friend.

Scheduling interviews is unavoidable, but be sure to leave extra time between appointments. As a neophyte, I lined up back-to-back interviews, with just enough time to get from one to the next. A 10:00 A.M. appointment in one place was followed by another just three hours later in a town sixty miles away. I found myself racing cross country, unable to savor the geography between. I now leave extra time between interviews and set aside appointment-free days to explore. Predictably, some of the most valuable material is the result of unscheduled interviews or a spur-of-the-moment decision to turn right instead of left.

A small, pocket-sized microrecorder has replaced my pen and pad. I discovered I was unable to transcribe quickly and accurately and soon realized people disliked repeating themselves. Initially, I tried using a larger cassette recorder, but found people tensed up and often couldn't take their eyes off the device. Tucked in my shirt pocket, with its microphone exposed, the small unit is unobtrusive. After pointing it out at the beginning of discussions and asking permission to record comments, I find that people usually forget about its presence. When finished, interviewees dictate their name, address, and telephone number into the recorder for later reference, should clarification be required. An address also makes it possible for an author to be certain that those who have helped are notified when the book becomes available. I send an autographed, complimentary copy to people who have been especially helpful.

Whatever the season, a camera is my constant companion, and picture taking a critical component of my fieldwork. Photography is an excellent tool for learning more about a place. Research for each of my full-color, book-length regional geographies included taking several thousand slides. Photography serves several purposes. It forces me to stop, look, and think. When trying to capture something on film, one is forced to look with a more critical eye. Specifics and relationships that might otherwise be passed over in a visual sweep of a scene are highlighted in the viewfinder and recorded on film.

Picture taking also produces a collection of visual images that can be drawn upon during writing. Short of revisiting places and people, projecting slides provides a means of jogging memories and rekindling emotions. Since a good part of regional geography for the general public has to do with description, pictures can help assure vividness and accuracy.

Photography can result in pictures suitable for inclusion in the finished product. Unless you are an accomplished photographer, however, you should be prepared for a bruised ego when "superb" images are passed over by the publisher. Many people with a 35-mm. camera assume they take professional quality photos. But pictures that work well in a journal article or garner praise when projected in a classroom or on the living room wall at home may be unsuitable for inclusion in a popular publication. Photo editors and graphic designers who demand professional standards of subject matter, composition, sharpness, and lighting often reject slides that an amateur would think merit enlargements.

Good photos can contribute as much to a popular regional geography as good writing. As well as illustrating geographic aspects, photos also show familiar scenes, sites, and landmarks, reinforcing the perception among locals that the book is about their own special corner of the world.

One of the challenges facing a photogeographer is not only to illustrate, but also to photograph the ordinary in such a way as to cause residents to look anew at the familiar landscapes around them. If photographed appropriately, pictures of bare hills, agricultural fields, small towns, cityscapes, and other subjects evoke new or heightened emotion among readers. To help achieve that goal I pay close attention to my own emotions when photographing and try to figure out what moves me. My first day in the field usually yields mere snapshots that look as though they were shot by a seven-year-old with a disposable cardboard camera. It isn't until the second day that something clicks other than the camera and I am able to begin viewing and recording scenes with a more artistic and interpretive eye.

WRITING FOR AN AUDIENCE

Although I am aware of the advantages of allowing research and writing to influence one another more directly by doing some of each simultaneously, my writing still tends to follow research. When I am ready to begin writing, the preliminary organizational outline that launched the research usually has been modified greatly. Topics and balance may have changed and entire sections been added or deleted depending on research results. Knowing the desired length of the finished manuscript, I assign general word counts to each section. This is a critical step often skipped by authors, who then berate editors who are forced to cull portions that may represent weeks of work.

Hoping for a good start, I always begin by writing the section I assume will be easiest. It might deal with a familiar topic relating to a personal research effort or a subject I found especially interesting. It need not be the first section. In fact, I find it best to write introductions last.

In no special order and in brainstorming fashion, I next list facts, points I want to make, phrases, quotes from interviews, observations, titles of articles, impressions, photos, and anything else I want to consider for inclusion in this section. Once material is ordered around a general section outline, I commence writing the first of what usually is two drafts. Using this part-by-part approach, I skip back and forth through the manuscript outline making sure not to leave all the "hard stuff" till the end.

Polished second drafts of sections are then read aloud as I listen. It is amazing how repetitious words, split infinitives, awkward sentences, lack of clarity, and unnecessary words surface when written prose is spoken. I am convinced this step is a prerequisite to good writing.

Having heard all second drafts, I begin considering the manuscript in a more holistic sense, while rewriting, editing, and working on transitions between sections. The polished third draft is read aloud, this time from beginning to end and preferably in one sitting. Once necessary changes are made it is ready for the publisher.

For most mortals, occasional writer's block is an unavoidable affliction brought on by lengthy writing projects. Recognizing and accepting the necessity of multiple drafts helps, but most writers experience times when they can sit in front of a keyboard for hours and not compose a single coherent sentence. Cures, when possible, are individual. For some a brisk walk or a change of scenery are all that is required. Others must let the condition run its course. As a high school student in suburban Detroit in the early 1960s, I discovered that the then-new Motown sound of the Supremes did wonders for me if played at 95 decibels. A quarter-century later, Diana, Mary, and Flo still are my salvation.

Popular regional geography may be dominated by information and descriptions, but it need not be boring. If the subject matter has been selected and research conducted with readership in mind, good writing—crisp, clean, clear, and inviting—can provide the link between author and audience. And good expository geography is much the same as other forms of popular prose. Authors who are new to popular geography can benefit from reading works of others. Those serious about writing for the lay reader may also want to invest in several good "how-to" writing books and consider enrolling in a creative writing course.

For me, writing for the general public is the most difficult of compositions. It is hard work. The challenge, and special reward, is to create reader interest and enjoyment. Readership of a popular book is not captive. Unlike students in a required college class, this audience is free to drop out at any time. An author of popular geography must continually ask, "Are my readers still with me?" Once a writer earns widespread recognition and advances to the leading edge of popular regional writing, perhaps, concern for readers at the conscious level can slide and emphasis shift to a less encumbered and higher level of artistic expression.

For academics, the most formidable obstacle may be an inability to shift from a guarded and controlled scholarly writing style to a freer and more open style necessary for good popular prose. An author must be willing to let curiosity, enthusiasm, and personality show through. Without those qualities, the voice readers need to accompany them through the writing is lacking.

Substantial detail, on which many geographers thrive, is one of the surest ways to dampen reader interest and lose an audience. Readers may be able to plow through confused and uninspired writing, but excess detail and statistics are certain to stop them in their tracks. When extra detail is essential, consider using "sidebars." They allow an author to present a case study or other specifics that might otherwise interrupt the flow of the main body of text. The public is familiar with journalistic use of sidebars and generally has no problem skipping over one without feeling excluded by the author.

This is not to suggest that technical matters need be excluded, but they must be presented in a way that does not overwhelm an audience. I have found that readers enjoy moderately challenging sections where opportunities for learning may be greatest. The intelligence of lay readers never ceases to amaze me; it would be a fatal mistake to underestimate their sophistication.

People seem especially interested in their local physical landscape. A topic such as bedrock geology can be dreadfully dull or can be written with the public in mind, as the following example illustrates:

> It was a collision, of sorts, that terminated the geologic tranquility of Western Montana. The first rumblings of change date from approximately 170 million years ago when the Paleo-Pacific plate and the North American plate converged just to the west of Montana. Since the crustal material of the oceanic plate was more dense, it buckled under the western, continental edge of the North American plate. Like the front end of an automobile after a head-on collision, the edge of the overriding continental plate was shortened, crumpled and lifted upward. This marked the beginning of the formation of Montana's first Rockies (Alwin 1983, 12).

And geomorphology becomes palatable when discussed in terms of the familiar and endeared:

> Montanans probably are more familiar with the swarm of laccoliths just north of the Adels. Here, in the apex between the Sun and Missouri rivers, a cluster of more than a half dozen intrusives rises from the plains. Among them are Cascade Butte, Haystack Butte, Crown Butte and Shaw Butte, each a landmark. The area moved Charles Russell, Montana's noted western artist, who chose it as a setting for many of his Old West paintings. Flat-topped and symmetrical Square Butte, seven miles south of Fort Shaw, is recognizable in the background of several of Russell's works. For him, the landscape captured the mood of an earlier era when range riders or Blackfeet in battle array might be seen on the horizon (Alwin 1982, 15).

Popular writing must be free of disciplinary jargon or other words that leave the average reader floundering and reaching for a dictionary with the turn of almost every phrase. For authors with impressive vocabularies, this is not the place to flaunt their lexical mastery. To do so not only is rude but also constitutes poor composition, since clear communication is the prime objective of any writing.

Use of a thesaurus is indispensable and is not cheating. But rather than look for more *im*pressive words, look for more *ex*pressive ones. Select words and phrases not only to describe and report, but also to convey mood, image, and emotion.

In the following passage, I was attempting to inform readers about rural depopulation in eastern Montana in the 1910s, but I also wanted them to visualize the exodus:

> The dry Depression decade of the 1930s witnessed a continued thinning of rural populace in many areas. Today, older residents pensively recall what became almost a wholesale abandonment of some farming areas. Many recollect the late Teens when there seemed to be a family on almost every half section, and four or five neighbors might be seen from a farmhouse window. In the '30s they watched settlement thin when up to half a county's farms might have folded. By the thousands once hopeful farmers simultaneously were pushed off the land and out of the region by impossible weather and economics, and were pulled away by the hope of jobs both near and far. Most who left never returned (Alwin 1982, 60).

To present statistics on the population, lumber production, university enrollment, and wholesale trade of Missoula, Montana, not only would have put readers to sleep, but would have failed to capture the city I know:

> Plaid-shirted loggers, millworkers with lunch buckets, university intellectuals, businessmen in three-piece suits, hold-over hippies and more contemporary "granola people"—Missoula has them all. Butte may claim title to Montana's most mixed population of nationalities, but no community in the state comes close to Missoula's conglomeration of politics, philosophies and occupations. Missoula is a collage of free spirits, a pot that refuses to melt, and one of Montana's most interesting and stimulating cities (Alwin 1983, 104).

Pick up a copy of even our most respected geography journals or scholarly books, and note sentence structure. In many cases simple and repetitive sentence construction dominates. When asked to evaluate manuscripts, one of the first things I do is a "*The* Count." I skim over the composition noting how many paragraphs start with the word *the*. If it is a prevalent first word, I brace myself for what probably will be a rigid read. When closer inspection reveals stretches with as many as seven and eight consecutive such sentences, I known it is going to be painful. Monotonous sentence structure will not work for readers who are expected to remain conscious.

Many of the same scholarly works by geographers are characterized by mega-paragraphs of 400 to 500 words. Such massive walls of print are visually too intimidating for lay readers. Limiting all paragraphs to one or two sentences is going too far in the other direction, but there are many options between one- and fifty-sentence paragraphs. And, again, variety is the key.

Reader interest and enjoyment and a sense that yours is their geography can be enhanced by periodically keying your text to the familiar. For example, distribution of forest cover west and south of Spokane, Washington, can be discussed strictly in terms of "here to there" or can be put in a more experiential context:

> Metropolitan Spokane is largely "in the trees" and a 200-square-mile prong of ponderosa carries the forest almost 30 miles farther onto the basalt plain. Travellers heading west out of Spokane via I-90 don't break out of the trees until about 25 miles southwest of the city, and those driving toward Pullman on well-travelled Highway 195 don't see the southern edge of the pines in their rear-view mirrors until approaching Spangle (Alwin 1984, 48).

Kittitas Valley, Washington, residents may yawn at another cattle inventory, but can smile at themselves when their Old West image is placed in a more personal context:

> No place in Washington is so dominated by cattle and nowhere else in the Evergreen State is the cowboy complex so well-developed. Each Labor Day weekend the Kittitas hosts one of the nation's major rodeos. Year-round, valley residents, on a per capita basis, wear more western belt buckles and cowboy boots than in any other place in the state. People don't look twice here if young intendeds sport cowboy hats in their engagement pictures in the Ellensburg *Daily Record,* or if a candidate for local office poses for campaign posters wearing a Stetson (Alwin 1984, 89).

Use of occasional short quotes garnered from interviews in the field can animate popular prose. There is no point in straining to say something in your own words when a local already has provided a "gem":

> [Columbia Basin] Project grape acreage has grown with interest in wine grapes region-wide, but still accounts for little more than one percent of total crop value. "It's a loud crop. There's a lot more noise about those grapes than acres," says Les Gius, who owns a 60-acre Red Delicious orchard west of Quincy. "In comparison to apples, the wine grapes are peanuts," he says (Alwin 1984, 99).

Finally, geographers occasionally need to remind themselves that popular geography should be fun to read. To help accomplish that goal I rely on a little "ooh, aah" and periodic levity. Statistics on the colossal Pleistocene Spokane Floods in eastern Washington are well suited to "ooh, aah" treatment:

> Geologists now know that the ice dam holding back Lake Missoula periodically burst sending as much as 500 cubic miles of water roaring through the Clark Fork Valley toward Eastern Washington, probably exceeding the legal speed limit on today's high-

ways along the route! Calculated peak flow rates of 750 million cubic feet per second (about 15 cubic miles per hour) would have been about 20 times the combined flows of all the world's rivers. Chances are maximum flows lasted only a few hours, with the draining of the lake taking perhaps a week or two (Alwin 1984, 26).

While describing a western Montana identity, I couldn't resist commenting on regional attire:

> It's not that Western Montanans don't dress up. Some businessmen, for example, wear three-piece suits to work every day. But chances are in most settings they wouldn't stand out if they didn't. Even when well-turned-out, a Western Montanan isn't bedecked in the latest Paris or New York fashion. Our tastes for styles seem to run a little behind many other parts of the country. Certainly other Western Montanans have sensed that they aren't quite up to date when they step off a plane at Denver's Stapleton International or Seattle's SeaTac and feel as if they have just passed through some sort of fashion time-warp (Alwin 1983, 8).

The personal rewards and potential benefits of popular geography to society and the discipline are great. But until populist geography is legitimized, be prepared for your efforts to carry little weight in tenure and promotion review. If they are denied credit for its research/creativity value, faculty members can make a strong case for recognition of popular geographic work as service to the public and the discipline. There are encouraging signs that our field is awakening to the importance of communicating directly with the public.[5] If you're interested, don't wait for an official seal of approval. The time is right, and people in "your" region are waiting!

REFERENCES

Abler, R. F. 1987. What shall we say? To whom shall we speak? *Annals of the Association of American Geographers* 77:511–24.

Alwin, J. A. 1982. *Eastern Montana: A portrait of the land and its people*. Montana Geographic Series, Number 2. Helena, MT: American Geographic Publishing.

———. 1983. *Western Montana: A portrait of the land and its people*. Montana Geographic Series, Number 5. Helena, MT: American Geographic Publishing.

———. 1984. *Between the mountains: A portrait of Eastern Washington*. Northwest Geographer Series, Number 1. Bozeman, MT: Northwest Panorama Publishing.

Borchert, J. R. 1987. *America's northern heartland: An economic and historical geography of the upper Midwest*. Minneapolis: University of Minnesota Press.

Garreau, J. 1981. *The nine nations of North America*. Boston: Houghton Mifflin.

Hart, J. F. 1982. The highest form of the geographer's art. *Annals of the Association of American Geographers* 72:1–29.

Kidron, M., and Segal, R. 1987. *The new state of the world atlas*. New York: Simon & Schuster.

Lewis, P. 1985. Beyond description. *Annals of the Association of American Geographers* 75:465–77.

Moon, W. L. H. 1982. *Blue highways*. Boston and Toronto: Little, Brown and Company.

[5]The nation's first National Geography Awareness Week in 1987, the AAG's annual John Brinckerhoff Jackson Prize for popular books in American human geography, and even the inclusion of the "Press Clippings" section in the *AAG Newsletter* are recent examples.

The Writing of Scholarly Books in Geography

John A. Jakle

How to write a book? This is the line of inquiry that I have been asked to advance. As appropriate as that question is to this anthology, it would nonetheless be pretentious for me to write directly to the topic. I am far from mastering the art of book writing. Thus I will focus on a related topic: why an author, especially a geographer, might undertake to create a book. I will offer some arguments, pro and con, on the scholarly tome as a communication device in academic geography and present some observations on book writing and the book publishing process. My remarks, like those of the other essayists in this volume, are highly personal, being grounded in my own adventures and misadventures in publishing.

GEOGRAPHY'S LACK OF BOOKS

Why might a geographer write a scholarly book? The question seems easily answered. Wouldn't all scholars want to publish their research in book form? The fact is, however, that North American geography has a long tradition of rejecting books as inappropriate vehicles of communication, especially for reporting research. The long article published in a major journal, such as the *Annals* of the Association of American Geographers, has been the favored outlet. There, one's work is certain to come to the immediate attention of geographers and to stand as a kind of permanent record signifying geography's best. Scholarly books, supposedly, should be reserved to senior scholars capable of synthesizing and extending long-term research endeavors; besides, much of their work has previously been published in article form and thus already subjected to peer review. Only definitive tomes should be encouraged to stand as monuments to personal (and thus, indirectly, to disciplinary) achievement. Apparently, only one area of book publishing has been deemed wholly appropriate for all geographers: textbook writing.

The result of this thinking has been to make geographers underrepresented on the nation's bookshelves. While we wait for our experienced scholars to produce their great works, relatively few great works actually appear. Some senior geographers do not enjoy a long enough life. Others grow tired and suffer physical and emotional exhaustion before producing the *magnum opus*. Others abide by traditional behavior and continue to craft articles, perhaps out of habit. Some attempt the "great book" only to fail, perhaps for lack of practice. Creating a book is very different from creating an article. It requires grappling with subject matter on a much broader scale, and usually to a more penetrating level. Writing successful articles does not guarantee an ability to write a successful book even when a scholar is mature and wise. Consequently, what academics in other disciplines—or, for that matter, lay persons—see of geographers in bookstores and libraries is a mixed bag: the few scholarly books that younger geographers have published despite spoken and unspoken taboos, the very few great books by well-established practitioners, and lots of textbooks.

Scholars outside geography tend not to read our journals unless in search of specific citations reflective of narrow interests. The selective mining of journals can hardly be expected to give nongeographers a comprehensive feel for the discipline. Not only may nongeographers have an incomplete view of geography, but geographers themselves, lacking synthesizing works at the book scale, may suffer a diminished view of the discipline as well. Books, after all, are not important as monuments to ego so much as they are important as communication devices, part of the symbolic interaction that forms the discipline as a society of scholars. To the extent that certain kinds of communication are underrepresented, a discipline cannot help but be diminished. For geographers, one outcome may be a general lack of confidence in themselves. To what extent do geographers suffer an inferiority complex regarding their discipline and assume, by way of defense, a perceived need to be better than scholars in cognate fields? To what extent is the traditional hesitation to expose oneself prematurely in book form a reflection of that attitude? Even discounting such speculation, one important question remains: What do we publish that readily demonstrates our worth both to ourselves and to others? Can article and textbook publishing suffice to demonstrate our value?

We have in geography a less than fully developed literary tradition. There is little emphasis in graduate education on writing skills. We tend not to grow intellectually by thinking of ourselves as gifted storytellers imaginative at word selection, phrasing, rhythm, and emphasis. We tend not to look upon our writing as literary or artistic. Rather, we write much as we would speak, as if literature were merely the casual spoken word written down. That few geographers are elegant writers goes without saying. Elegance is hard to come by. That most geographers are yet unaware of the importance of attempting elegance needs frequent repeating.

Certain intellectual stances in geography can be identified as diverting attention away from good writing. First, geographers have tended to emphasize exploration over journalism. That is, discovery has been emphasized over reporting. Finding out has been primary; the telling of one's discoveries secondary. (Of course, the best geographers—certainly the best-known geographers—do both with equal flair.) Second, geographers have emphasized science over art, that is, method over communication. Once an appropriate research design has been achieved, then the reporting of analytical results is allowed to follow directly. Too many geographers report facts in the order of their discovery as befits a demonstration of proper method. Too few seek to create intellectual excitement through innovative thought construction. Third, geographers have been overly reliant upon maps, and upon statistical and other quantitative devices. Too often the search for precision and objectivity has turned geographers away from imaginative word use and sentence structure. "Good writing" is too often confused with abstruse writing—a supposed sign of intellectual achievement. In addition, we depend too heavily on our graphic and numerical displays to communicate our sense of science and discovery. Although these traditional stances are appropriate to a modern discipline rooted in scientism, they may offer inadequate encouragement to good writing and book production.

Times are changing, and many traditional views and stances are on the wane. Never before have North American geographers been such active book writers. The increased social relevancy of the discipline's traditional subject matter certainly invites geographers to share their ideas with wider audiences than ever before. Never before have scholars in other disciplines been as interested in questions of spatial organization and location, human spatial behavior, landscape, and place. Indeed, so many practitioners in other disciplines are "doing geography" that geographers had better enter the book-writing fray with vigor or risk losing whatever proprietorship they possess over traditional scholarly orientations. Academic geography must work at remaining a principal marketplace for ideas traditionally considered "geographic." Book writing will be critical to that end.

Such disciplinary needs may or may not fit the career trajectory of the young scholar. A geographer should take care in establishing early publishing goals by debating the pros and cons of article versus book writing. Geographers do tend to read professional journals, especially the major serials that appear in mailboxes with regular frequency. Articles will come to the immediate attention of peers, even those outside one's narrow area of interest. Article reprints mailed to colleagues serve as a kind of public relations device. Books, on the other hand, may not be noticed immediately even by those who share one's interest. Books are expensive to give away in self-promotion. Yet journal articles become "obsolete" quickly because of their limited breadth or depth. They easily become lost in the back issues of journals shelved and forgotten. They are retrieved only with effort by nonsubscribers, especially nonacademics. Books, on the other hand, enjoy a longer life, if only for their wider breadth or greater depth. They are easily retrieved in libraries. A book has a separate presence on a library shelf; it is an obvious package. Articles, as dictated by journal editorial policies, are necessarily limited by restrictions placed on length, format, subject matter, and orientation. Authors are variously limited regarding the scale of problem attacked and the extent of elaboration allowed. Although books do not permit unfettered license, the degree of freedom can be substantially broader. Usually, a book author is able to experiment.

Authors who might want to promote their work beyond geography will find their own professional journals of little help. Geography journals are relatively few and their circulations limited. Geographers publish in the journals of other fields not only to increase their visibility beyond geography, but to give themselves added credence at home. Such publishing, however, deprives the geography marketplace. Scholarly book publishing, by contrast, is a way whereby geographers can broaden an audience while, at the same time, not detracting from academic geography as an intellectual emporium.

What kinds of books might geographers write? Up to this point my remarks have been aimed primarily at research reporting, writing rooted in scientism. Ideally, such scholarship involves a clear statement of problem through the posing of specific hypotheses set in a review of relevant literature. It details data collection, verification, and assessment. Standard techniques of data analysis are outlined, or innovative analytical strategies detailed. Conclusions are derived from hypothesis testing and placed in clear theoretical context by way of interpretation and evaluation. The essay, however, is another kind of presentation. Rooted more in the humanities, it serves as a topical review. It is a synthesis of diverse observations built around general as opposed to specific hypotheses, a kind of explanation sketch. From an obvious starting point, the ideal essay proceeds logically from observation to observation toward a significant conclusion that stands as greater than the sum of its contributory parts. Great books may merge both kinds of writing.

Some books stand more as pilot studies, others as definitive works. Most lie somewhere in between. Pilot studies can serve legitimate purposes and should be viewed as more than merely preliminary or tentative. Ideally, they are exploratory, striking out across new ground using new points of view. They are suggestive of categories of concern appropriately to be developed in the future. Pilot studies invite constructive criticism. Their purpose is to excite interest, if not controversy. Pilot studies call attention to new opportunities even where established scholars are content with old ways. The definitive work takes a very different stance. It is well anchored in established scholarship, since major ideas have already been well tested. Conclusions are substantive, embracing wide reach and deep penetration. The definitive work serves as a capstone to research of long dedication, rather than as a stepping stone along an intended path.

Books of various kinds are appropriate and necessary to a discipline. Ultimately, a field's reputation is anchored on research essays of the definitive sort. Yet, to discourage other or "lesser" kinds of books would be to diminish the intensity of that mutual sharing that supports and sustains great scholarship. Foreground buildings may represent important architecture, but without the context of background structures even the best architecture suffers. A skyline of only a few very tall buildings is not a skyline at all. Lesser buildings are necessary to provide a sense of mass. So it is with an academic discipline. Monuments are important to an intellectual landscape, but without other elements in place they lack meaning.

GETTING MANUSCRIPTS REVIEWED

The modern themes of academic geography are not the subjects that publishers traditionally seek. The largest corporate publishing houses and the most prestigious university presses have made their reputations publishing other subjects. They have traditionally courted nonfiction authors who are topically oriented—i.e., authors who focus on specific subject matter in either contemporary or historical contexts. Few acquisitions editors have backgrounds in geography. Thus, geography manuscripts may be rejected as not conforming to better-known disciplinary templates, or accepted only to the extent that they complement familiar disciplinary approaches. Publisher book lists in geography are few. Most lists are oriented to topics, questions, and methods of analysis around which other disciplines are oriented. Thus, even if accepted, a geographer's work may find itself "orphaned" in a publisher's offerings with few or even no like companions. Such books are harder to promote and to sell.

Just as academic geography is not well known to editors, so geographers are not well known either. Only a few scholars may be visible enough in a geography subdiscipline to qualify as manuscript referees. Thus a would-be author may encounter the same referees, with the same narrow range of opinions, over and over again. A manuscript may be judged more on the basis of politics than on its merit or potential for revision. These "gatekeepers" may reject work that does not conform to their own inclinations as scholars. Hypocrisy too frequently enters such decisions. That is, reviewers apply higher standards in assessing the work of others than in assessing their own work or the work of their close colleagues. Constructive criticism is withheld and a would-be author left with little more than simple rejection. Such reviewers "safeguard" the field against competing work, points of view, and analytical orientations that would diminish their own options. Young scholars should not be surprised by such treatment. They should expect to work hard at obtaining a publisher by selling both themselves and their discipline. They must expect to confront negative skepticism and convert it to constructive criticism by constantly seeking clarification.

Geographers form into cliques and tend not to be supportive of those beyond their group. Academic geography in North America is something of a political party composed of exclusive cells. Perhaps the best way for a young geographer to get published is to have a politically astute mentor who carefully guides research and aids in placing work. True excellence often results from such relationships, for the weight of experience is brought firmly to bear with cross-generational implications. But not all of us have had mentors. And some of us have interests that look beyond a current generation's preoccupations. Authors who are poorly connected should not despair. They simply will have to exert extra effort to have their work objectively assessed, since it will not have been "pre-adapted" by weight of connections. Geographers may like to think of the discipline as totally objective and want non-geographers to think so also. However, the field is merely a society of human beings. It may seek objectivity like other academic disciplines, but, as in other disciplines, biases exist rooted in social realities.

Egotistical referees will especially antagonize would-be authors. These are the sort of people who feel obligated to demonstrate their own intellectual worth by denigrating work under review. Such people act, at least in part, to boost their own egos at the expense of colleagues. Some have never published books themselves and do not want to see others do so. Authors learn to recognize such behavior. By comparing referee evaluations with assessments made by book reviewers after publication, one begins to gauge one's work objectively.

Of course, all authors should seek reviews of their work before submitting a manuscript to an editor. Such peer assessment will not only eliminate gross errors and omissions but also alert authors to the effectiveness of their tone and orientation. Sometimes authors are bound too closely to the details of their project to place it in context and effectively capture reader interest. Unfortunately, it has been my experience that one cannot always depend on colleagues to help, especially when approaching those outside one's immediate network of acquaintances. It is important to cultivate a network of well-placed colleagues. Geographers may find scholars in cognate disciplines more cooperative and seek their help in bringing manuscripts to maturation. This is especially true of those authors who hope to conceptualize and execute work outside the traditional mainstream of their sub-discipline.

LAUNCHING A BOOK

For most authors, book writing holds a special personal challenge. Projects are launched not because a book fits a sense of disciplinary propriety, but because it promises self-fulfillment. This sense of fulfillment goes well beyond the mere pursuit of career. One starts with an idea and appraises it in terms of previous publications and what one envisions of ongoing research and writing. Then one comes to some decision as to the book's intended objectives and audience. One can proceed by developing drafts of various chapters and using helpful colleagues to advise on revision. With a mature manuscript in hand, one may approach a publisher through a letter of inquiry or even through direct unsolicited submission of the manuscript itself. It is, perhaps, wiser to proceed more cautiously—to engage various publishers right from the beginning of a project's conceptualization and, indeed, to involve them in conceptualization.

A number of potential publishers should be identified. Their appropriateness depends on what they have already published and are continuing to publish. Would a geography title fit comfortably into a publisher's list? Examine the list, evaluating the kinds of books by subject and looking critically at the nature and quality of the scholarship evident. Evaluate carefully the format and physical quality of those books most similar to the planned project. Are the paper and binding of high quality? Are illustrations clear? Look especially at editorial details. How evident are typographical and other errors? Several guides exist to aid would-be authors in identifying potential publishers. The guides provide brief company histories, identify editors, and describe the kinds of manuscripts sought (Hubbard and Bachman 1988; Neff 1988; R. R. Bowker Co. 1987). None of this information, however, can replace careful firsthand appraisal of a publisher's products, and even inspection of the facility. Best of all, talk to others who have previously published with the same publisher.

A detailed proposal with outline, a c.v., and, perhaps, even a demonstration chapter should be sent to several publishers that have been deemed appropriate. The purpose of these informal submissions is only to excite interest. Interested acquisitions editors will send these materials to referees who, later, may also be asked to review the finished manuscript if formally submitted. An enthusiastic editor may offer a potential author a contract. Such contracts might best be left unsigned, because they only give the publisher an option to publish. An author is committed to send the publisher the manuscript if finished, but the publisher is not obligated to accept it. The publisher will want to obtain referee criticisms. An unscrupulous publisher might stall a manuscript's review, publish a competing title in the interim, and ultimately reject a project and leave an author unpublished. The cash advance, of course, represents one reason why authors might want to place themselves under contract. An advance on royalties may be necessary for fieldwork, preparing graphics, or otherwise completing a project. A cash advance signals a publisher's intense interest.

Word processors have vastly speeded the writing of manuscripts and, more importantly, their editorial revision. A first draft should, nonetheless, be allowed to season. After some elapsed time it should be scrutinized and modified as appropriate. A second draft should be shared with colleagues who are willing to assist. All criticisms should be appraised carefully; authors must also appraise what is not being said. Often it is not the content of a manuscript that elicits negative response, but its omissions. Something may be neglected that even critics may not see directly, but only sense subconsciously. The author will discount or reject much criticism, but should assess all comments carefully.

THE PUBLISHER'S RESPONSIBILITIES

What should an author expect from a publisher? Few acquisitions editors will have a manuscript reviewed if it is being considered by another publisher. Many authors withhold that information, but most, I suspect, are honest in their dealings. When an editor holds a manuscript for an undue time without responding, then an author may be justified in starting the review process elsewhere. Usually, an editor's referees remain anonymous to encourage greater honesty and more useful criticism. An editor may solicit a list of potential reviewers from an author. This may be as much a test of an author's awareness as an attempt to identify the most appropriate critics; therefore, one should always have referees in mind when submitting a manuscript. Duplicate copies of a manuscript may speed simultaneous reviews, although many publishers deliberately seek reviewers in sequence, asking authors, in turn, to comment at each stage. Two or three reviews will usually be sought. When reviews are favorable, most editors will synthesize reviewer comments and suggest to the author a series of modifications. Once the author responds, the entire package (manuscript, reviews, and author responses) will be taken to a senior editor or, as is usually the case at university presses, to an editorial board. If the decision is made to publish the book, and a contract has not already been signed, a contract will then be offered.

Authors should read contracts carefully. Some may contain clauses that commit future work to the same publisher's initial review or that restrict the author from publishing on similar topics with other publishers. Authors should make sure that royalty arrangements are clearly understood. In setting royalties a publisher is determining the physical quality of a book, the number of copies to be printed, and the price. Authors need to discuss these issues fully, even to the extent of appending typed memoranda of understanding to the printed contract form. The highest royalty rate may not work to the best advantage of either the publisher or the author. The higher the royalty, the higher the price—with fewer books sold and lower profits. Relatively few books published today contain large numbers of maps or illustrations. A publisher may not be expert at handling visual materials, and a contract may need to be appended to cover graphics.

Once a final manuscript has been received, most publishers require at least a year to bring a book to market. The author will be asked early to complete a questionnaire geared to planning a marketing campaign. Copy editing will begin, and eventually the publisher will ask the author to approve changes in wording, sentence structure, and organization. This is usually the author's last opportunity to consider extensive modification, given the high costs of type resetting. Several months after returning an edited manuscript, an author will see galley proofs (the text set in type) or even page proofs (the text set in type and arranged in page format). Sometimes page proofs contain illustrations, but often they do not. The author's prime task at this stage is to check that type has been properly set and that words, sentences, and paragraphs have not been omitted inadvertently or placed out of proper sequence. Ideally, an author should also see "blue lines," the corrected page proofs with illustrations inserted. Unfortunately, soaring book production costs have led many publishers to omit the blue-line stage, thus depriving authors of the opportunity to catch new errors introduced with the previous resetting of type. The foreshortening of the editing process has contributed substantially to the epidemic increase in typographical errors in recent years. Finally, with page proofs in hand, the author may be asked to create an index.

Most publishers promise to bring books into print within a reasonable period of time, and to promote them through normal advertising and marketing channels. The author's contract will also specify the conditions under which a book will be allowed to go "out of print." Usually, the publisher will arrange for the author's copyright. But the author should fully understand his or her rights to buy remaining copies and even printer's plates, should the publisher allow the title to lapse. Changes in the tax laws now make it unprofitable for publishers to warehouse large numbers of books. Thus, first issues tend to be smaller, and books are kept in print for much shorter periods than formerly. Many nonprofit presses, particularly university presses, seek only to recover costs with small first printings. If a book is successful, profits will come from second and subsequent editions.

BOOK REVIEWS

Books are the stuff of scholarly interaction, and book reviews bring that interaction to the fore. Although quantifiable measures (such as sales volume or citation indices) attest to a book's success, it is from the book reviews that authors most immediately and most directly sense the prospective impact of their work. Although there is a long tradition of book reviewing in geography, compared to some other disciplines our journals carry relatively few reviews. Journals also give little emphasis to review articles, which compare several works in assessing large questions. In addition, many geography journals refuse to print commentaries related to already published book reviews. Thus authors have no redress when their works are unfairly treated by reviewers—save through the courts. The discipline is thus denied a dialogue that can clarify issues and stimulate thought. Sloppy or unprofessional reviewing goes unchallenged.

Just as authors have an obligation to bring to print the best possible products that they can manage, reviewers have a similar obligation. Necessary qualities for book reviewing include: mature comprehension of the subject at hand, careful reading and evaluation of the book being assessed, and fairness of judgment. Unfortunately, too many reviews in scholarly journals, those of geography included, are capricious. Too many geographers abuse their review privileges. They fail to identify clearly a book's purpose, sketch its contents, or outline its conclusions; they may even distort its contents. Too often evaluations are overly self-serving, using the review as a platform to advocate a cause. Some reviewers use reviews to pursue personal vendettas. Some may not review a book at all but, through sarcasm, deliberate exaggeration, or like devices, review instead the book that they might have written or would have liked to have seen written. Some reviewers write to curry favor with authors, or withhold criticism so as not to generate disaffection. Reviews that intemperately attack a book or, conversely, heap undue praise upon it usually contain a hidden agenda.

All authors deserve objective evaluations of their work. Reviews should analyze books according to stated objectives, the means applied, and the conclusions reached. Weaknesses should be pointed out, but strengths should not be ignored. The criticism should be constructive in that it enables, if not encourages, the author or other scholars to refine and advance the line of work. A book's weaknesses and strengths should be kept in context and not removed, isolated, belittled, or championed as a reviewer's mood dictates. At stake is not only an author's psyche, but a discipline's integrity. Authors undeservingly discouraged may withdraw, potentially depriving the field of intellectual energy. Authors overly encouraged may not extend themselves in the future, also depriving the field.

PROMOTING A BOOK

Book reviewing is important business. Reviews are a kind of promotion, since they either encourage or discourage readers from engaging a work firsthand. Few geographers have time to read extensively. Even the most avid readers must be selective, and for most the book review is the most important screening device. Reviews also influence what librarians purchase, and thus a book's availability. Authors should expect that their work will be evaluated fairly and hope that it will be acclaimed critically. Some authors seek to do more than hope: they cite, quote, or praise potential reviewers frequently in their work to foster a climate of admiration. They may encourage sympathetic colleagues to volunteer reviews to book review editors. They may work through acquaintanceship networks to stage-manage a positive welcome for a new title. Articles published and papers read at meetings may be little more than advance publicity for an impending work. Authors tempted by such strategies should tread carefully lest they slip into the quicksand of unprofessionalism. In promoting a book, geographers should avoid becoming mere hucksters.

The best way to promote a book is to craft it carefully. One might engage other minds in conceptualizing the project. One might seek extensive criticism at various stages of a book's evolution, not just at the second-draft stage. The objectives are to anticipate criticism and to adjust the narrative and illustrations to communicate advocated points of view most effectively. One does not have to abide by every-critic's suggestions, but one should meet those suggestions head on to sharpen one's advocacy. In this regard the preface is very important. Many readers will not get beyond the preface; it is there that interest is either aroused or lost. Many reviewers will rely solely on a book's preface for primary insights, so this is where authors should state succinctly their purpose and avenue of approach. They should place their work in context, elaborating its significance, and hint at their conclusions. Above all, the preface should anticipate potential criticisms.

Marketing questionnaires are never fun to complete. At the end of a project an author may be wrung dry of enthusiasm. With research, analysis, and writing completed, many authors want only a change of direction. Other projects now beckon the imagination. Nonetheless, authors should work carefully to compile lists of journals and newspapers where their book might be reviewed. They must think carefully about what makes a book distinctive, important, and valuable, so that the publisher can produce informative and effective advertising copy. If a popular audience is targeted, the author should stand ready to answer reporters' questions and engage in radio and television interviews.

CRAFTSMANSHIP AND CAREER

Most publishers have guides that instruct potential authors on the mechanisms of book writing: how to structure footnotes, how to integrate illustrations, how to construct indices, and so forth. Proofreading instructions, complete with a guide to proofreaders' marks, are usually included in these guides. Special care should be taken to avoid libel, a topic emphasized in most guides. Libel is anything that accuses another person unjustly of doing something unlawful, disgraceful, or ridiculous, thereby holding that person up to public ridicule. Libel may result from direct accusation, but it can also result from strong innuendo. The author's task is to keep others from thinking that they have been libeled. If criticism is to be leveled, it must be framed objectively, without undue emotion. It should be gauged to enlighten and not to defame. It should be fair and not serve primarily to boost the writer's ego through a display of intemperance.

It is helpful to know the rudiments of copyright law—to know what is and what is not in the public domain. Care should be taken to avoid plagiarism. Decisions on what to quote, where to quote, and how to quote vary with the type of exposition and the author's objective. But giving full credit where credit is due is important, even in instances where paraphrasing has been extensive. One should hesitate to paraphrase one's own previous writings and seek to bring only fresh material to the fore. Many observations may bear repeating across the spectrum of one's total work, but they should be rendered each time with renewed imagination. One should watch the length of a manuscript, making it as concise as possible. Publishing is expensive, and a publisher may turn down a project simply because the potential product is too long. A project might be broken into several books, or residual pieces might be published as articles.

Some writers have sought to develop a personal literary style appropriate to the kind of writing they do. Most fall into a style, however, without giving the matter much thought. Others deliberately experiment, trying different formulas to suit different circumstances. Some geographers carefully fit their books into career trajectories, with each book leading toward and thus contributing to others. Some pursue a single topic, exploring it across various geographical contexts; others pursue different topics as they relate to a single place or region. The author of several closely related works may achieve success as a scholar more rapidly than others, since the weight of scholarship falling upon a limited set of issues makes the overall contribution more readily evident. Careful thought should be given to one's scholarly image as reflected in the style of writing and sequencing of books.

Why then would a geographer want to write a scholarly book? Why submit oneself to the trials and tribulations of unsympathetic or self-serving reviewers, not to mention the emotional sting of honest, constructive criticism? Why shoulder the logistical burdens of manuscript preparation and promotion even where publishers proceed with dispatch and flair? Why bother with a kind of publishing not universally encouraged by one's peers? Beyond the potential benefits to geography, the reasons ultimately remain personal. Certainly a book-writing tradition, especially a highly literate one, can only help strengthen the field. Down deep, however, it is the personal drive in each of us that counts. It is the sense of accomplishment that derives from having conceptualized and completed a project at the book scale. It is the satisfaction of having met a challenge above and beyond the ordinary and of experiencing growth and maturation as a result. Beyond the missionary zeal with which we justify our research and publishing, a sense of personal fulfillment propels us all. To the extent that publishing scholarly books enters the picture, we should be responsive, if only to be honest to ourselves. Geography is what we make of it. And what we make of it will reflect the extent to which honest, personal drives are given seed and nurtured. If the idea of writing a book suits us, we should cultivate it.

REFERENCES

Hubbard, L. S., and Bachmann, T. M. eds. 1988. *Publishers directory*. Detroit: Gale Research Co.

Neff, G. T. 1988. *Writer's market: Where to sell what you write*. Cincinnati: Writer's Digest Books.

R. R. Bowker Co. 1987. *Literary market place: The directory of American book publishing*. New York: R. R. Bowker Co.

Writing a Textbook

H. J. de Blij

Textbooks have a special place in North American higher education. In universities elsewhere in the world, textbooks often are recommended as supplementary reading, but rarely are assigned for compulsory study or as specific sources for answers to examination questions. Worldwide, the textbook-as-study-guide phenomenon, if it exists at all, occurs at the first-year (freshman) level. When I began my studies as a geography student in Africa, the introductory course was team-taught. Professor J. H. Wellington occasionally referred to a decades-old British textbook, but often urged us to "go to the library and look this up for yourselves." Professor S. Jackson referred to papers he had written for journals. "They are a bit technical," he would say, "but see what you can make of them." Other lecturers did much the same. By the end of the year-long course, we all had a list of books to which reference had been made. Some of these were available in local bookstores; others were not. The library was indispensable.

When I arrived in 1956 at Northwestern University in Evanston, Illinois, to begin a Ph.D. program, I was amazed that a series of classroom courses would be required, and astounded that each course had one principal textbook that each student was supposed to acquire. I complained about this to the chairman of the department, Professor C. F. Jones, the famed Latin Americanist. It was a humbling experience. With a few sharp questions, substantive and theoretical, Jones proved that a textbook—any textbook—would teach me much about the realm covered by his course, for which I had been instructed to register. By the end of the academic quarter I had bought every geography textbook on Middle and South America I could afford, including Jones's own classic (Jones 1930).

The notion that a textbook could be an integral and vital part of a university course was strengthened further during my years as a graduate teaching assistant in the large introductory physical geography class taught by Professor W. F. Powers at Northwestern. (Those were the days when this was a respected three-quarter sequence that fulfilled college science requirements, and before the "new geography" zealots destroyed it and doomed the department.) In his memorable lectures, Powers managed to complement, and not to recapitulate the textbook (Finch and

Trewartha 1949). In combination, lecture and text provided a view of the field unmatched in my experience.

Not every encounter of this kind proved so satisfactory. Courses on Africa and political geography were accompanied by books that I felt were not worthy of the classes; and one particular course was poorly served by a book that did not represent the discipline well. Still, even the inadequate textbooks were of interest. They provided an insight, I believed, into the way the courses in question were taught by prominent geographers at distant universities.

I sustained my interest in textbooks throughout graduate school, but I also realized that textbooks were not held in high esteem as academic contributions by colleagues who would make decisions regarding promotion and tenure. So, during my first faculty appointments, I worked on articles and monographs rather than textbooks. But, following examples I had admired in graduate school, I prepared detailed and carefully organized outlines for every lecture I presented in each assigned course. At the time I did not anticipate it, but those course and class outlines later became the bases for several textbooks.

COURSE AND TEXT

It is axiomatic that the functions of textbooks change as the course level does. The basic information needed in a freshman geography text ought to be less essential in a senior-level book. A textbook to accompany a senior-level course on Africa, for example, should be written under the assumption that fundamentals such as climatic, biotic, and demographic distributions would form part of students' mental maps. Or so I thought when I wrote *A Geography of Subsaharan Africa* (de Blij 1964); the preface mentions differences between this and preceding books of its kind. First, there is a justification for confining the book to Africa south of the Sahara, the geographic realm of Africa. Second, an allusion is made to a "thematic" approach: "This book does not attempt to be factually complete. . .not every irrigation scheme and hydroelectric plant appears in the index. . .not every large city is mentioned" (de Blij 1964, iv). Third, a sample is given of country-topic associations, which form the themes of the chapters: the land question in Kenya, federalism in Nigeria, and so on.

Reaction to these innovations was, to put it in the most favorable light, mixed. Both reviewers and users of the textbook complained that the basics of African geography were missing and that the thematic approach deprived lecturers of the opportunity to pursue these themes themselves. Reports from the field, however, indicated that the sometimes tendentious thematic issues did contribute to considerable classroom discussion and debate. But the absence of "basics" inhibited sales of the book. The publisher wanted me to revise it to approach more closely (i.e., to compete more directly with) other available texts. But I could not contradict the purpose of the first edition in a second one. Not until 1977 did a successor appear (Best and de Blij 1977).

From this experience comes a lesson of consequence for any would-be textbook author: the interests and objectives of writers and publishers can easily diverge, particularly in the smaller markets (i.e., the advanced-level book). Two decades ago, publishers—especially major publishers—were prepared to publish marginally profitable books as a matter of prestige and as a signal to the profession that they valued its business. Specialized, small-circulation books were part of the lists of several major publishers. Today, the costs of production and the demand for greater profits keep many good books of this kind from publishers' geography lists.

Innovation, therefore, has its risks—it can be too much of a good thing. That proved to be the case with a junior-level (third-year) book designed for political geography courses (de Blij 1967). My idea was to combine substantive text with original readings from the journal literature. Having taught political geography for several years, I knew that far too few students used the library to read the articles I recommended in conjunction with the text I was using. Thus I created a book that made available both the basics of the field and the seminal journal literature. In this case, the first edition was widely adopted, and the second edition (1973) enjoyed even stronger acceptance. But then skyrocketing permissions fees began to interfere with the concept of the book. When the third edition appeared, authored by M. Glassner (Glassner and de Blij 1980), the expenses involved in the inclusion of articles were so great that other costs of production had to be cut back, as is evident from the appearance of that edition. By the fourth edition (1989), the inclusion of journal articles had become prohibitively expensive, resulting in a book now far removed from its original concept.

There are times when a textbook can have a major, positive impact on the course of the discipline. Such was the case with the first appearance of *Geography: Regions and Concepts* (de Blij 1971). I had taught the large introductory world regional geography course at Michigan State University for years, and had become convinced that the course had merit as an introduction to geography—but only if fundamental theories, concepts, and ideas in the field were placed in regional context. This was a time when colleagues from many leading institutions were assuring me at every annual meeting of the Association of American Geographers (AAG) that regional geography would soon be a dead issue, and that regional courses—such as those on Africa, South Asia, and Middle America—would soon follow it into oblivion. Some of these colleagues were actively engaged in the destruction of such departmental offerings, on the grounds that geography, as a science of universals, could not dabble in regions and their qualities. In 1967, the noted geography editor at John Wiley & Sons, Paul A. Lee, suggested that I write a proposal for a world regional geography book, in spite of the fact that such courses were declining and even seemed moribund. I wrote the proposal, part of which became the preface to the 1971 book, and Lee managed to persuade the publisher that regional geography would survive and, in future years, revive.

After three years of work and more than 2,400 pages of manuscript, *Geography: Regions and Concepts* was published in what must be the most unappealing format possible: a bulky, gray mass of print with poorly reproduced photographs and barely readable maps. But the book's impact was unmistakable. The notion of combining regional settings with geographic concepts gave new life to world regional geography courses. I received an unprecedented flood of mail from colleagues, administrators, and even students (many of whom complained that the book was too long and difficult). Acceptance of the book far exceeded the publisher's expectations, and after an unsuccessful attempt at condensation (de Blij

1974), it was published in full color in 1978—the first textbook of its kind to be so endowed.

In my experience, therefore, a successful and original course with a well-developed outline forms the best basis for a good textbook. Few faculty members teaching geography courses are satisfied with the assigned text; there always are opportunities for improvement. These may take the form of substitute handouts; in turn, the supplementary material might coalesce into drafts of chapters. And thus a textbook might be born.

AUTHOR AND PUBLISHER

When a faculty member decides to write and produce a textbook, a relationship with an established publishing company usually follows. It may begin through the overture of an editor who has heard of the author's plans; or the author may write to an editor, suggesting the development of a particular type of text. With these first steps begins an association fraught with problems and even dangers. The faculty member views the book as a contribution to the discipline, a step forward, a way to help leave behind the old and tired. The publisher knows that too much innovation can limit the sales potential of a book. The author, especially in the case of an advanced-level textbook, is in it, substantially, for the prestige and visibility a good book brings—even a textbook. Today more than ever, the publisher is in it for the money.

The possibilities of misunderstanding are almost endless, and they are best mitigated (if not avoided) by the author providing a lengthy, clear, detailed statement of intent, a complete outline of the proposed book, and a sample chapter. The statement of intent should cover the author's estimate of the potential market (and its possible expansion as a result of the book's contribution), the existing "competition," and the level at which the book is aimed.

It is important that this statement about the proposed book also make clear to the publisher exactly what the author expects in terms of production. Good geography textbooks tend to include various kinds of illustrations, especially maps and diagrams. Publishers want to see these illustrations submitted in final form as part of the manuscript—"camera ready," as the favorite saying goes. Authors, on the other hand, want publishers to invest in cartography and preferably to prepare it at the publisher's expense. The statement of intent should be clear on this matter, so that the author's wishes become part of the contract. Otherwise, a misunderstanding can begin that may later ruin the project.

The statement should further include a writing schedule. Every editor has bought lunches and dinners to ensure the commitment of the author, has had to justify these expenses to the company, and has never received the manuscript. The author should agree that a failure to meet the proposed writing schedule (barring possible circumstances on which author and publisher should agree in advance) voids the contract. Thus, would-be textbook authors ought to reflect on their past performance. Was the dissertation produced on schedule? Have previous writing commitments been met on schedule? Were galleys and page proofs returned to journal editors as instructed? If the answer to any of these questions is negative, would-be authors should do themselves and the publisher a favor by declining the project. The textbook business is a business. Time, to publishers, is money, and unmet commitments are costly.

A promising proposal is likely to attract the interest of more than one publisher. Editors will send your statement to colleagues who serve as editorial reviewers; news of the project will quickly diffuse. Responses may range from enthusiasm ("will absolutely adopt this book in my course if it is published") to denigration, none of which should influence the author. Many reviewers are not sufficiently well informed to render useful judgments, but many publishers demand that editors secure such reactions. Assuming a generally positive response, one or more editors may present the author with a contract for the proposed work.

From the textbook author's point of view, the contract is a sequence of clauses that might be better identified as pitfalls. I have never personally involved an attorney in the matter, but lawyer friends who have read several of my contracts have described them as "not worth the paper they're printed on—except from the publisher's viewpoint." That may be an exaggeration, but the contract offered by the publisher generally has the effect of binding the author while leaving the publisher free to terminate it. For example, every contract in my file has a clause that reads, approximately, as follows: "Following receipt of a manuscript deemed acceptable by the publisher, [the company] will publish the work." What the words "deemed acceptable" mean is that if the publisher changes its corporate mind, it need only judge the final manuscript as "unacceptable," and the publisher is no longer bound to publish it.

In another standard clause, the author is bound "not to publish, or cause to be published, any book that competes with the [said work] while this contract is in effect." But the publisher does not agree not to publish one or more competing works. This relates directly to a further clause that stipulates that "artwork prepared for [said book] is and will remain the property of the publisher and may be used by the publisher in any way it decides." As a result of this combination of clauses, a publisher is in a position to use the original ideas represented on company-drawn maps (or other illustrations) in books that compete directly with the one you have just proposed. When authors raise questions about this practice, the standard corporate answer is that the competing book is "aimed at a different segment of the market."

Prospective textbook authors sometimes concentrate their attention unduly on what appears to them to be the key issue in the contract: the royalty clause. In general terms, the geography market is not large; even at the introductory level, sales are small compared to most other fields. In my experience, textbook authors in geography have inflated expectations regarding the returns on their time and effort. The quality of production, the amount and quality of cartography and other illustrations, and related aspects of the contract are far more important than 1 or 2 percent on the royalty rate. In any case, the royalty rate has become a quagmire. Royalties are paid on the number of copies sold. A couple of decades ago the rate was a percentage of the list price, and authors (or their auditors) could calculate their royalty income fairly accurately. Today, royalties tend to be paid on the "net income" of the publisher, which can mean so many things that only an unaffordable team of experts could trace the actual flow of funds.

With very few exceptions, textbook authors should not expect their incomes to be enhanced substantially by royalty payments. I recently calculated the time spent on (and income yielded by) a senior-level text that for several years was the leading book used in a particular course. The book required some 740 hours to write, not including preparatory work, of which I kept no detailed record but which perhaps required 200 hours. Checking copy editing and reading galleys and page proofs required 95 hours. Preparing the index took 65 hours. I paid for the cartography both because I wanted to keep control of it and because the publisher would provide only a small subvention; the cost was $4,500. The publisher printed 7,000 copies, some hundreds of which were given away as complimentary and review copies. I received an average of $1.71 per copy in royalties, for a total taxable income over the life of the edition of about $11,500. Deduct author's production costs, and my income amounted to $5.97 per hour.

Thus the royalty clause, and trading other advantages for 1 or 2 percent extra on royalty rates, is far less important than it might seem. At the introductory (freshman) level the situation is naturally different. Where tens of thousands of copies may be sold, a 1 percent increase in royalty rates can make a substantial difference in an author's income. But publishers' options at the freshman level also are greater.

One additional issue of the contract merits attention: permissions. In the past, publishers routinely supported authors' needs for items requiring permission (usually for a fee) from other publishers for reproduction, such as extended quotations, certain classical illustrations, and photographs. More recently, authors have seen the costs of such permissions deducted from their royalties. When the contract is negotiated, the publisher should be asked at least to share the cost of a reasonable quantity of permissions.

To summarize, textbook authors should seek to mitigate the disadvantages with which they are confronted when preparing to approve a publisher's contract. This can be done by adding clauses relating to cartography, illustrations, and other ancillaries. Royalties currently amount to 10 to 15 percent of net, except in unusual circumstances; the royalty rate is less important than other clauses, some of which will be discussed later. And authors should temper their income expectations from textbooks.

WRITING AND PRODUCTION

For many authors, including myself, the most difficult aspect of writing a textbook is getting down to it, clearing the desk and devoting time that could be spent so much more pleasantly (and, in most cases, more productively). Writing a textbook demands constant attention to the needs and limitations of the intended readership. It is an exercise in education as well as communication and is quite different from any other form of scholarly writing. Changing gears from writing articles or monographs to textbook preparation is a difficult challenge.

In a sense, textbook writing entails professional risks. It is the ultimate exposure of weaknesses as well as strengths: explain something awkwardly or, worse, incorrectly, and hundreds of colleagues will know about it. No matter how balanced and comprehensive, a textbook reveals one's preferences and predilections. There is no hiding behind esoteric specialization. A textbook is a revealing personal statement, providing insight into the author's teaching, explanation, organization, and perception of the field. There will be those ready to debate the book's premises, its pedagogy, and its contribution to the discipline. There are geographers who have written dozens of articles in journals but who are known and discussed principally for one or two major textbooks. Textbooks can reveal; they also can obscure.

In the best circumstances, the writing phase should involve a close and continuous cooperation between author and editor. The author should adhere to the writing schedule; the editor should prepare a review schedule to match the author's efficiency. Unfortunately, the ideal situation—in which a design is prepared before the book is written and then becomes an integral part of the writing phase—almost never happens. It is advisable to prepare the cartography in sketch form before the text is written; having other illustrations in hand as well is even better. Most publishers, unhappily, wait until the text is "in production" before the "photo search" begins. By then it is too late to integrate photos and other illustrations effectively into the text, as could otherwise have been the case. But the cartography, all by itself, can hold up the book's production. Getting publishers started on maps at an early stage is a bit like trying to push a glacier.

As the writing progresses, the editor becomes the liaison between author and publisher. Following revision, the manuscript is "put into production," as the publisher's jargon goes. This may not happen, again depending on the publisher's rules, until every last page of revised manuscript has been received. This, of course, delays the progress of the manuscript to finished book by months, and fortunately some sensible publishers have changed that rule.

Part of the practice of delay has to do with a concern that the author will not revise the chapters as the reviewers recommend. In fact, authors should use textbook reviews in ways very different from scholarly book reviews. Reviews of scholarly works are in the same category as, say, peer reviews for an AAG *Annals* article. They may be addressed point by point. Reviews of textbook chapters should be treated (and written) differently. The most useful reviews of textual material deal with errors, possible misinterpretations, recent (perhaps unreported) developments, and additions and deletions of material. Often, in my experience, reviewers' recommendations cancel each other out. One reviewer may insist that a certain chapter should be expanded; another may suggest that it be condensed. The author, and not the editor or the publisher, should decide what to use from the reviews the editor provides.

Production, once it begins, can be a harrowing experience. Several books of mine have been virtually destroyed by incompetent copy editors and only partially repaired. In the second edition of *Regions* (de Blij 1978), for example, the old Jeffersonian notion of the "primate city" was introduced; once familiar, the term was used in the context of old European and then-new colonial capitals. A copy editor eliminated all references to primate cities, and wrote a pedantic note in the margin that "as a professor, you should know that 'primate' refers only to apes and bishops." I located all her deletions and reinstated them; when she received the checked manuscript she angrily telephoned me about the matter. I sent her a copy of the first page of Jefferson's article and reported the matter to my editor, certain that this copy editor would never again be employed by so prestigious a publisher. I was wrong.

The copy-editing problem is a pernicious one. Authors of textbooks (and, for that matter, authors in general) should demand, if necessary as part of their contract, a sample of the work of the copy editor who will edit their manuscripts. Many copy editors today do not know the fundamentals of English grammar; this seems no longer to be a requirement for the job. True, the author always can reverse editorial errors. But given the tight return schedule for galleys and pages demanded by the production department, that may not be feasible.

Another item of consequence in textbooks is the index. Some textbooks require both a glossary and an index, and terms for the glossary should be marked as the manuscript is written. The index, however, often is prepared in haste and, as a result, inadequately. Some publishers will agree to have the index prepared by their staff or a freelance indexer. This is convenient and sometimes necessary if the author's schedule demands it. But the best indexer for a textbook is the author. Freelance indexers (especially those who routinely handle "science" book indexes) are not sufficiently familiar with geography, and as a result the indices of some geography texts reveal amazing omissions.

Sometime during the early production stage, authors will receive "sample pages" representing the design of their book. These sample layouts should contain every item included in the manuscript: text (with all levels of headings represented), "boxes," maps, other illustrations, bibliography, glossary, and so forth. Authors should respond carefully to the publisher's queries, and express their own views on the proposed layout. In textbooks the choice and size of type, the width of columns, the amount of white space, and the general visual appeal of the format are very important. Some publishers do not understand how significant the relationship is between headline size and the author's organization of the material. Bold first-level heads should be differentiated clearly from less prominent second-level heads and still less bold third-level heads.

Perhaps two months after the index has been submitted and checked, the book will appear, and the effort to sell it now becomes the publisher's priority. Authors may expect prominent advertising in major journals, and may be disappointed to see very little of this. In fact, page advertising in the professional literature sells few textbooks. If this form of advertising has a purpose, it is to make the name of the author more familiar, in general terms, than it may already be; but, as a sales device, it is not cost-efficient. Direct-mail advertising to teachers who might adopt the book in their courses is more rewarding, but authors should be prepared to cringe at the content of some of the mailing pieces prepared by publishers' sales departments. These are not the staid announcements that may follow the publication of a scholarly book or monograph. Unless the issue is covered in a clause in the contract, making advertising pieces subject to author's approval, authors should expect to be surprised—not always pleasantly!

A scholarly book or monograph stands by itself, while a textbook is likely to have appendages, especially at the undergraduate level. For books in large introductory courses (but also in such standard upper-level courses as economic or urban geography), publishers like to have student study guides, instructor's manuals, and even slides and overheads available to "assist" students and faculty (read "encourage adoptions"). These ancillary materials constitute a considerable additional task for the author. The study guide must contain useful exercises and hints to help students comprehend the material in the text; this may require additional maps and diagrams. The instructor's manual includes suggestions to enhance teaching effectiveness and perhaps textual material to supplement what is in the book; the instructor will expect the manual to contain examinations (in objective and other forms) for every segment of the book.

The preparation of these materials should be planned at the beginning of the project, not as an afterthought. The publisher should invest in this work, either by providing a nonrecoverable grant to the author, or by supporting someone else to work on the development of ancillary materials from manuscript and galleys, as the book evolves.

Student understanding is much enhanced when certain illustrations (for example, maps) in the text can be projected on the screen by the lecturer during classes. The provision of a film roll of the crucial maps and diagrams is an inexpensive and attractive proposition. The author can take the initiative in the creation of this service to his or her colleagues. Of course, it should be expected that some lecturers will use the film roll from one book while adopting another!

REVISIONS AND EDITIONS

Unlike scholarly books and monographs, textbooks are subject to frequent revision if they have a long and successful life. Revisions are needed because the discipline progresses; textbook material must be kept up to date. In recent times the used-book market has played a growing role in publishers' decisions to prepare new editions. The publisher makes a major investment in the creation of a book, invests hundreds of thousands of dollars in purchasing paper and press facilities, and risks much in binding and storing a large stock. Then a book is sold, over and over again, after being sold new only once. Profits made by purveyors of used books exceed those of the author and publisher. This situation has much to do with the rise in bookstore prices of textbooks, and the ultimate losers, of course, are the students.

At present there is nothing to protect the originator of the book, the author (although relief is sure to come). In the meantime, textbooks often are placed on a "revision cycle" that may form part of the original contract. Some basic books are on three- or four-year cycles; advanced-level books may be revised at longer intervals.

Unfortunately, the temptation is strong to keep costs down by changing format rather than content, doing a minimum of updating, and letting appearance prevail over substance in short-term revisions. In fact, frequent revision (not to mention slower cycles) presents opportunities to innovate, to improve, and to set the field forward. But the job is massive and comes close to that required for an altogether new book. The publisher may require a totally new manuscript, and the design and production process begins again from the start. I keep a file for every segment of any book subject to revision that includes articles, comments from colleagues, letters from students, clippings, and so forth. I use these files when the revision begins.

There is no such thing as a "revised" printing, contrary to what is sometimes surmised. It would be impractical to change population statistics and data in the text because the new figures would not match the unchanged cartography. A second and third printing of a certain edition can be used to correct minor errors that were not caught in galleys or page proofs. But a revision, no matter how early, is a new book. An author who agrees to sustain a revision cycle makes a major commitment of time for the future.

JOINT AUTHORSHIP

Textbooks frequently are coauthored. It is a matter of record that multiple-author books (in which as many as a dozen individual authors each write one or more chapters) tend not to do well; quality is uneven, approaches differ, and schedules are met with varying reliability. But a number of two-author texts have benefited from the infusion of new ideas and energies. One of the most prominent is the team of Strahler and Strahler. The senior Strahler made an enormous contribution to the survival and status of physical geography when the first edition of his well-known text appeared (Strahler 1962). Since the 1970s, his books have been coauthored (Strahler and Strahler 1978). For decades the Strahler texts maintained an originality and currency that kept them in the forefront of a growing list of introductory works in this field.

Other examples come readily to mind. Joint authorship has obvious and proven advantages: fresh viewpoints, complementary capacities and experiences, and accelerated writing schedules. But there are pitfalls. A publisher's contract, signed by both authors, says a great deal about the responsibilities of the authors to the publisher (and, as noted, rather less about the publisher's commitments to the authors). Significantly, the contract says nothing about the responsibilities of multiple authors to each other. Many a promising textbook project has been nullified by the failure of one author in a partnership.

Occasionally the author of a successful textbook will invite a colleague to join the venture. In such instances the coauthorship situation is rather different, because the book already exists, has achieved its niche in the market, and is the product of the lead author's initiative. The advantages of introducing another author include the infusion of new energies and ideas, a broadening of perspectives, and perhaps a widening of available expertise in fields covered by the book. However, potential disadvantages exist. There is a difference between revision by reconstruction and revision by addition. The former is a much more demanding task, but the latter shows more clearly the involvement of the coauthor. There is thus a temptation, particularly on the part of junior authors, to make their mark on a book by expanding it. This is not always improper, but the size of a book is a factor in production costs as well as market acceptance. Another potential problem lies in ideological reorientation. After the publication of a new edition of one of my books that had been capably revised by a colleague, I received several letters from unhappy "users" complaining that the book was now "too liberal" and prescriptive for their taste (I had enthusiastically endorsed the coauthor's work). Acceptance of the book declined, but over time its changed posture won a new audience and the book regained its position.

In the development of any textbook under joint authorship, the authors should prepare a statement of responsibility that covers all aspects of the project, as assigned to all parties. One of its clauses should deal with the issue of publisher's advances, both on royalties and for developmental purposes. The days of large advances against royalties are generally over, but publishers like to cement their relationships with authors by offering token amounts (usually $500 to $1,000) to each member of a writing "team." The joint-authorship agreement should include a stipulation that repayment will be made if an author fails to meet his or her obligations. Such a negotiation may appear unnecessary in the first months of enthusiastic cooperation among authors who also may be close personal friends. But the years of writing and production will bring stresses severe enough at times to break the strongest bonds of friendship and collegiality. When that happens, a written agreement can make the difference between temporary and permanent damage.

TEXTBOOKS AS SCHOLARSHIP

Textbooks are routinely differentiated from scholarly books (as has been done in this essay) because their objectives are commercial as well as educational and their content is deemed a summation, and not an advancement of knowledge. In some universities, textbooks work against, rather than for, the academic careers of their authors as far as promotion and tenure are concerned.

The prominence of a successful textbook may obscure other professional contributions made by its author. In addition to my textbooks, I have written and edited eight other books, three of which were published by university presses. But soon after the appearance of the first edition of *Geography: Regions and Concepts,* my identity as a textbook writer was entrenched. When I discussed my experience with other geographers who have continued to publish research as well as textbooks, it became clear that this was no isolated instance. Nor are all the connotations negative. A widely adopted textbook generates a network of connections among colleagues and students, and a continuous flow of communication, most of it constructive and enlivening. It remains true that attitudes toward successful textbook writers range from envy and disapproval to admiration and support, but the positive consequences far outweigh the negative, notwithstanding the low esteem in which textbooks are sometimes held.

Textbooks do, however, contribute importantly to scholarship. They establish the status of a field in the discipline, and record its state of knowledge. They indicate the salient themes that drive the field. They outline the directions the field is taking. They convey the significant contributions made by leading scholars, thus connecting the research literature to the introductory course. They demonstrate the methods by which research findings are made and promulgated. I have seen many a textbook segment that, as a piece of scholarship, exceeded much of what appears in the profession's peer-reviewed literature.

Another important aspect of a successful textbook is the impact it can have on the field it represents. The appearance of such a book can stimulate the introduction of courses where previously none was offered. Good textbooks can contribute to the reform of curricula. Furthermore, they can help educate college and university administrators who, having been schooled during the 1960s and 1970s, often suffer from the geographic illiteracy that is now so endemic.

Finally, a good textbook invites competition and, inevitably, some imitation. None of this should distress authors; if a publisher is willing to invest in a field already well served by a good textbook, this can only contribute to expansion, diversification, and growth. The nurturing of a discipline is an objective shared by all and best achieved by demonstration and example. When it comes to good textbooks, the more the better.

REFERENCES

Best, A. C. G., and de Blij, H. J. 1977. *African Survey.* New York: John Wiley & Sons.

de Blij, H. J. 1964. *A Geography of Subsaharan Africa.* Chicago: Rand McNally.

———. 1967. *Systematic Political Geography.* New York: John Wiley & Sons. 2nd ed., 1973. 3rd ed. (with M. I. Glassner), 1980. 4th ed. (with M. I. Glassner), 1989.

———. 1974. *Essentials of Geography.* New York: John Wiley & Sons.

———. 1971. *Geography: Regions and Concepts.* New York: John Wiley & Sons. 2nd ed., 1978. 3rd ed., 1981. 4th ed. (with P. O. Muller), 1985. 5th ed. (with P. O. Muller), 1988.

Finch, V. C., and Trewartha, G. T. 1936. *Elements of Geography.* New York: McGraw-Hill, 1936. 2nd ed. 1942., 3rd ed., 1949.

Glassner, M. I., and de Blij, H. J. 1980. *Systematic Political Geography.* 3rd ed. New York: John Wiley & Sons.

Jones, C. F. 1930. *South America.* New York: Holt.

Strahler, A. N. 1962. *Physical Geography.* New York: John Wiley & Sons.

Strahler, A. N., and Strahler, A. H. 1978. *Modern Physical Geography.* New York: John Wiley & Sons.

EDITORIAL AND GRANT-GETTING "SECRETS"

Not-So-Mysterious Secrets of Publishing Journal Articles

Bonnie Loyd

The key ingredient of a strong journal article is brilliant research. But we all know that. We begin grappling with research in graduate school and spend the rest of our careers trying to develop more sophistication in conducting studies. But although intelligent research is essential for a publishable manuscript, it is not the only requirement. As a journal editor, let me suggest a few practical ways to make a manuscript more alluring to editors and reviewers. During my graduate school education we seldom discussed the mechanics of publishing, yet now I find that authors who understand the process can make it work for them.

SELECTING A JOURNAL

Some authors undermine their prospects for publishing by completing their writing and only then scouting for a likely journal. That's a bit like making the wedding dress before you've seen the bride. These shortsighted types select a journal as an afterthought, and although they may make half-hearted efforts at adapting their manuscripts to a journal's style, they don't always succeed. In contrast to this myopic approach, I suggest that writers have two or three publications in mind when they begin their projects, so they can visualize at every step of the research and writing whether the final manuscript will fit comfortably into one of the journals. Authors may sometimes revise their plans about where to publish as they proceed with their work. That's to be expected. But I cringe when I sense an author trying to shove a completed "square" manuscript into our "round" journal. To ready a manuscript for publication, seasoned authors do the same thoughtful planning they do for their research.

Dozens of scholarly journals exist—far more than most young scholars realize. To find possible journals for your work, you might consult reference librarians, directories, indices, abstracts, and citations in other articles. When you are familiar with a range of publications, winnow your choices by first considering the audience of each journal. How large is the readership? Which scholars or decision-makers see the journal? Does the audience span several fields? Next assess the prestige of each publication. Scholars evaluate prestige with several measures, including how often articles from a journal are cited in other scholarly work, what proportion of manuscripts submitted are accepted for publication, and whether manuscripts are refereed anonymously by several scholars. Fortunately, B. L. Turner II (1988) as well as David Lee and Arthur Evans (1984) have published articles that elaborate on these points and rank journals, so some of the work of sorting publications in geography has been done for us. But all authors must consider the information in light of their own goals. Publishing an article in a small, little-known journal, for example, may be a sensible decision if its readers are exactly the people you want to reach.

PREPARING TO SUBMIT A MANUSCRIPT

The most valuable advice I can give authors is simple: know the journal thoroughly before you submit a manuscript. At *Landscape* we mail guidelines to authors that begin with the statement, "Our past issues are the best guide to our style and content, so we encourage you to review several recent issues." As you review, notice what topics the journal covers. Study how research is analyzed and presented in this publication. Calculate the length of articles. Consider how illustrations and tables are used. Because the first article in an issue is often one the editor has selected as the strongest, it may be an especially good model. Look most attentively at the work of scholars you admire. Be sure to inspect recent issues, because editors and editorial policies change in the volatile publishing world. Instructions for authors are usually available in the journal or by mail, so check for them.

Knowing the characteristics of a journal will guide you in tailoring a manuscript to suit it. At *Landscape* we regularly reject wonderful manuscripts simply because they don't fit our publication: they may be far too long, they may use quantitative analysis even though we prefer qualitative, or they may critique a building by a professional architect despite our focus on vernacular design. Most publishers and editorial boards have carefully defined their journal's niche in the publishing world after considering scholarly trends, competing journals, and economics, so they are unlikely to shift their focus for one article, no matter how splendid it may be.

As you begin writing, scrutinize the journal for specifics of presentation. Examine, for example, titles, abstracts, introductions, literature reviews, subheads, length of paragraphs, captions, and bibliographies. If you find, for instance, only one level of subheads in the journal, don't construct your manuscript with three levels. If literature reviews appear prominently in every article, don't submit your manuscript without one. True, no manuscript will be rejected simply because the subheads don't match the journal style. Reviewers will, however, be wary of a manuscript in which the author repeatedly violates obvious guidelines. They begin to suspect that the reasoning and the reporting of facts may be just as sloppy as the presentation. If a manuscript with rough edges in presentation is accepted, the editor and author will finally have to hammer it into conformity with the journal style, which is time-consuming and, most of all, irritating. Good editors and authors would rather cooperate to strengthen the writing and analysis.

To look like a professional when submitting a manuscript:

- □ Double-space everything in the manuscript, including quotations, footnotes, and bibliography, and also leave wide margins. Editors need space to make corrections and insert instructions for typesetters.
- □ Keep your name off everything except the title page, which the editor detaches before sending the manuscript out for review. This way, work can be reviewed anonymously. Don't give away your identity by including your name in captions, a biography, or footnotes referring to your previous work. You can add these bits after the manuscript is accepted.
- □ Use a shortened form of the article's title in the upper corner of each page, and be sure to number every page.
- □ Reproduce quotations exactly as they appear in the originals; spelling, punctuation, and capitalization must be identical. The only exception is the first letter of a quotation, which you may capitalize or lowercase to suit its position within the text. Keep a photocopy of each quotation to use in proofreading, and supply a copy to the editor if a quotation has unusual wording or style.
- □ Make the introduction represent the rest of your manuscript accurately. It should outline exactly what the paper covers, no more or less, and be exquisitely written.

For more detailed advice about preparing a manuscript, as well as pointers on language, consult John Fraser Hart's entertaining article, "Ruminations of a Dyspeptic Ex-editor" (1976).

All scholars submitting manuscripts to journals should be familiar with two points of scholarly ethics. First, in academic publishing it is never acceptable to submit a manuscript to more than one journal at the same time. Saving a few weeks might be tempting, but because editors and reviewers volunteer hours to evaluate and revise manuscripts, they are understandably annoyed if one is withdrawn to be published elsewhere. Once burned, editors may be reluctant to deal with the offending author again. Second, it is unprofessional to publish two articles describing the same research and conclusions or even to use the same paragraphs in two articles. Scholars can certainly cultivate several manuscripts from the same fieldwork, each developing a different theme and oriented to a different audience, but scholars who pad their lists of publications by duplicating their own work in several journals develop cloudy reputations. On these prickly issues Susan Hanson speaks with thoughtfulness and authority (1988). Her article presents an excellent survey of the subtle ethical questions in scholarly publishing.

Finally, when you submit manuscripts, don't expect editors to do your work for you. Specifically, don't send a 300-page thesis and ask which twenty pages might interest the editor unless you are *very* close friends. And, above all, don't submit a manuscript loaded with elementary errors in spelling and grammar and expect the editor to "fix it up."

Scholars should be armed with the finest reference books when they face the battle of writing an article. Dozens of guides fill bookstores and libraries, but as I spend more time in publishing, I find I rely on only a handful of old standbys. Nearly all publishers depend on the *Chicago Manual of Style,* and writers can too. It is the best single source on everything in publishing from footnotes to reading proofs. Although it addresses an array of publishing concerns, such as copyright law and typography, I consult it most often for advice on style, because the *Chicago Manual* sets forth standard ways to use punctuation, capitalization, abbreviations, numbers in print, and other specifics. When two reference books clash in their counsel, publishers nearly always abide by the *Chicago Manual.* If you are serious about becoming a more accomplished writer, you might consider reading this book in bed.

Writers and editors are frequently baffled about which dictionary to choose from the many choices. Some dictionaries include more slang, some offer more detailed etymologies, and some keep up with emerging vocabulary in science and technology. I have several dictionaries at home and at the office, and I'm always trying out more. However, many editors, including me, now use *Webster's Ninth New Collegiate Dictionary* as the standard reference, particularly for hyphens and compounds. As I edit a manuscript, I most frequently look up hyphenation. I wonder, for example, whether *housemate* is one word, two words, or hyphenated. I check both *Webster's Ninth* and the tables in the *Chicago Manual* for answers. Authors who use another reference, perhaps the *Random House Dictionary of the English Language,* may be puzzled about why my suggested hyphenation differs from theirs. Dictionaries vary on compounds and hyphenation because this part of the language is changing rapidly. Even though authors may have their own preferences about dictionaries, editors and proofreaders can't reasonably use different ones for each manuscript in a journal. So, unless your editor specifies another dictionary, I propose that you adopt *Webster's Ninth New Collegiate Dictionary.*

To move beyond details and learn more about clear, logical writing, I recommend, first, the famous book by Strunk and White, *Elements of Style* (1989). It is amazingly brief and valuable. Academic writers should also be acquainted with *The Modern Researcher* by Jacques Barzun and Henry F. Graff (1970). In addition to discussing clean prose, the authors address many issues scholars confront, such as defining a research problem, collecting appropriate material, and reasoning clearly. For more tips on language, I suggest any book by Theodore Bernstein of the *New York Times,* and *The Reader over Your Shoulder* by Robert Graves and Alan Hodge (1979).

USING PHOTOGRAPHS

I love photographs, and because so much of geography is visual, I'm surprised we don't see them more often in our publications. Photographs can illuminate articles with valuable information, and, as a bonus, they catch the attention of readers flipping through the pages. In some ways, though, preparing photographs for an academic journal differs from preparing photographs for a popular magazine such as *Life* or *National Geographic,* even if the basics of good photography are the same. As scholars we need to think primarily about presenting evidence rather than creating atmosphere or injecting human interest. I want my authors to be great documentary photographers, like the Farm Security Administration photographers of the 1930s, which means they should allow subjects to be seen directly, without imposing an obviously artistic point of view. But even within the boundaries of straightforward documentary photography lies an ample range of possibilities that few scholars explore.

When they assemble a collection of photographs for a journal article, thoughtful authors look for variety. They select, for example, close-ups and broad panoramas as well as the more common middle-distance views. Unfortunately, scholars who study architecture often drift into the habit of shooting every building from directly in front and about thirty feet away, creating a generic perspective. When their articles appear, readers are stupefied by bungalow after bungalow or bank after bank, all shown at the same angle and distance. Taking photographs from several distances, or with several lenses, both keep readers alert and add considerable information: panoramas show a building in its neighborhood context, and close-ups reveal construction techniques and materials—evidence missing in the standard middle-distance portraits of buildings.

Experienced photographers know many more strategies for increasing the variety, and thus the richness of information, in their illustrations. One effective but neglected approach is simply to incorporate people and activity into photographs. Regrettably, many photographs of street scenes appearing in research journals look as if they had been shot in ghost towns. These views are unappealing as well as misleading portrayals of places. Although a bit more time and more film may be required, authors can take photographs showing how landscapes are used by people and which people use them. Photography editors for commercial magazines claim that readers are fascinated by pictures of people, and this is true for scholarly publications as well. Why should we ignore this in scholarly publications? Authors can also enrich the variety in their photographs by exploring interiors along with exteriors, by trying vertical as well as horizontal compositions, and by experimenting with shots from rooftops, below, behind, or alongside. I encourage photographers to be imaginative, with the caution that journal readers want information, not gimmicks.

Geographers have a few special requirements to remember when taking photographs. Most of our illustrations show vegetation and buildings, and both can trip up amateur photographers. Leaves on trees, bushes, and lawns often blur in photographs, so you should set the focus and shutter speed carefully to ensure a sharp print. And if you record architecture frequently, consider investing in a PC, or perspective-correction, lens. The lens reduces convergence, which is particularly disturbing in shots of tall buildings. Also keep in mind that geographers like to see foreground. They want to see the soil, vegetation, or pavement in front of a bungalow, and you cheat them if you frame the bottom edge of the shot too tightly.

Although I find it difficult to convince authors, I know that color slides almost never reproduce well in black and white. They lose their crispness and contrast. In ten years of experimenting with conversions from color to black and white, I have seen only one batch survive well, and I'm still not sure what made that batch different. So, if at all possible, shoot black-and-white film in addition to color slides. Be especially conscientious about taking black-and-white shots if you are documenting a rapidly changing landscape or one in a remote location you cannot easily visit again.

As an editor, I think a lot about how to handle prints. To produce the best prints, we have negatives printed by a custom laboratory, not a drugstore service or speedy franchise operation. Authors should do the same. I ask authors not to frame images too tightly when shooting or printing photographs because we need leeway in cropping to fit a layout and, in addition, a small amount of the photograph is masked on all four edges when it is inserted into a flat by the printer. All authors should leave borders on prints. That way, prints can be handled by the borders, and layout artists can mark them for cropping. Borderless prints are for your relatives, not for your editor. If you must write on the back of prints, use a very soft pencil or marker. I prefer to number a photograph in the border with a special soft pencil and put other information on a separate sheet. Some clever authors write on overlays or on labels that stick to the backs of prints. In any case, don't let damp ink from the back of one photograph rub off on the front of another. Protect your photographs from bending, and never use paper clips on them. Experts recommend storing negatives and prints away from heat, light, and dust, preferably in archive-quality plastic sleeves that do not cause chemical damage.

Preparing photographs for submission requires the same attention to detail as preparing a manuscript. For most prints, authors should supply the name of the photographer or owner, the year it was taken, the location, and a complete caption. Editors can generally work with any size photograph and ask the printer to reduce or enlarge it to the exact size needed in the layout, but for the best reproduction it is better to reduce than enlarge. Although eight-by-ten-inch prints are glamorous, I find five-by-seven-inch photographs adequate for nearly everything except covers or double-page spreads. And even though I don't always succeed, I try to persuade authors not to overwhelm me with choices. If a journal can print six photographs with your article, don't send forty and ask the editor to choose. Even if the editor will spend the time, you may be unhappy with the selection. When you do send a few extra photographs, be sure to identify your preferred choices.

If you don't have original photographs for your article, you can often find illustrations from other sources, many of them free. Textbook publishers hire photo researchers to hunt for illustrations, but it's fairly easy to do even if you haven't tried before. *Picture Sources Four,* edited by Ernest H. Robl (1983), is the best place to start. This book lists nearly a thousand sources of photographs and indexes photographs by topic and geographical location. Many scholars also sift through photographs at historical societies, museums, government agencies, corporations, consulates, and tourist offices. The collections are often strong, and prints are sometimes free in exchange for a credit line and a copy of the published article. Still another tactic for finding illustrations is to track down the source of photographs used in other books on the same topic.

Although you may have paid for a photograph, this does not necessarily give you the right to reproduce it in an article. You must obtain written permission from the owner to publish any photograph or other illustration that is not your own, and sometimes you must pay an additional fee. The only exceptions are photographs, from government agencies and corporate publicity offices, clearly stamped, "For public use." See the *Chicago Manual of Style* for a sample letter requesting permission. You should supply your editor with a copy of the letter you receive granting permission. Customarily, authors are responsible for all illustration fees. Many sources waive fees or charge reduced rates for educational publications. Most sources request a credit line, and the proper etiquette is to use the exact wording they specify.

SURVIVING THE REVIEW PROCESS

Once you have submitted a manuscript, the editor initiates the review process. At serious scholarly journals, editors mail manuscripts to at least one, and perhaps two or three, experts on the topic. The reviewing is conducted blind—i.e., the reviewers do not know who wrote the manuscript, and the author does not know who is reviewing it. While the apprehensive author waits, the editor collects written comments from the reviewers and then conveys the final decision, normally eight to twelve weeks after the manuscript was submitted. Those are the bare outlines of peer review, but the process has subtleties well known to veterans.

In most cases reviewers agree whether a manuscript should be accepted or rejected; when they don't, the editor may make the final decision or send the disputed manuscript to still another reviewer. Some journals accept as few as 10 percent of the manuscripts they receive, while some take as many as 40 percent. At the major geography journals today, 30 percent is routine. If an article is rejected, the editor often sends photocopies of reviewers' comments or a summary of the comments to the author.

Frequently the editor responds to the anxious author with specific suggestions for revision rather than a straightforward yes or no. Some authors are puzzled by this and suspect it politely disguises a rejection. Quite the contrary. Editors are eager to help develop manuscripts into publishable articles, and after they invest considerable time in reviewing a manuscript, they want it to appear in their journal. Occasionally, editors must return drafts to authors two or three times for rewrites, but persistent authors have an excellent chance of being published. However, if you receive an outright rejection, don't resubmit your article to the same journal. Instead, revise it and try another publication.

The peer review process is intended to select the best manuscripts and select them fairly, but sometimes it doesn't work as smoothly as it should—a cranky reviewer may torpedo a sound manuscript that doesn't support some cherished bias. If you encounter an unreasonably persnickety reviewer, pull up your socks, incorporate the suggestions that make sense, and try another journal. All good manuscripts eventually find a home.

Every journal has its own variations on the review process, but the specifics of how it is done—the length of time it takes, the number of reviewers, and the acceptance rate—are usually available in reference books such as the *Directory of Publishing Opportunities in Journals and Periodicals* (1981), in guidelines for authors, or in annual reports prepared by the editors. Stanley Brunn has published an especially thorough article about reviewing (1988), which includes a list of common reasons for rejection.

WORKING WITH AN EDITOR

After a journal has accepted a manuscript, the author and editor begin another phase of the publishing process, working together to polish the article. They check facts, expand sections of the argument, eliminate repetition, reinforce themes, and improve the organization, as well as buff the writing style. If you learn how to tango with editors, they can help you transform your manuscripts into stronger pieces. If you view editors as adversaries, however, you may miss some valuable assistance. Just as scholars specialize in particular research topics, editors specialize in logic and communication. Accomplished authors take advantage of that expertise. Authors who trivialize an editor's work as mere fussing with commas probably don't understand what a good editor can do.

Of course, seeing your manuscript covered with changes can be daunting. Nearly every line may have been battered by the editor's pencil. But don't despair; keep in mind that heavy editing does not mean you are an incompetent writer, because even gifted writers have their manuscripts returned with abundant questions and revisions jotted in by editors. At the *New Yorker* magazine, for instance, editors spend months with authors, plowing through draft after draft, and most of the elite corps who publish in the magazine regard this extraordinary treatment as a luxury. But because scholars seldom receive such meticulous inspection of their prose, they may not know how to react to even superficial editing. When I compare notes with other editors, we usually agree that talented, experienced authors often respond graciously, even enthusiastically, to editing suggestions, while a few insecure, first-time authors bristle at every mark on the page.

The amount of attention editors devote to manuscripts varies widely. At some small academic journals, editors are faculty members with little time and meager backgrounds in copy editing, so they may send manuscripts directly to the typesetter without comment. But if you are lucky, you and your article will be embraced by a capable editor willing to spend a day or more poring over every line of your work. Perhaps no other reader will scrutinize your presentation so carefully.

A truly professional editor will do far more than correct spelling and punctuation. This wizard will point out gaps in the reasoning, suggest convincing examples, clarify conclusions, and rescue authors from embarrassing mistakes in names and dates. An editor's comments can stimulate a scholar to develop a more sophisticated analysis. True, not every editor works skillfully or conscientiously, so be prepared to protect your work from capricious changes. But don't be so defensive that you ignore thoughtful observations. Keep in mind, too, that editors have limited time and patience, and if they are exhausted by a torrent of routine spelling and grammar problems, they may never move on to more valuable tasks, such as examining the organization and content of your work.

Some scholars do not realize that good editing can be a cooperative process, with both the editor and author participating. Hence, the tango metaphor. The editor may lead, but the author must know how to follow gracefully, or the pair will only lurch across the floor. Few editors demand that all their revisions be accepted without question. In fact, responsible editors ask an author to review the edited manuscript and correct or improve the editing wherever necessary. Although this assignment sounds straightforward, many scholars stumble over it. A handful of contentious writers, for example, dispute nearly every whit of editing, usually in barely legible handwriting squeezed into the margins. They frequently justify their preferred phrasing simply by saying, "It *sounds* right to me"—an easy but inadequate rationale. Sometimes a careless author insists on a particular word, claiming it implies a whole complicated situation that requires a full paragraph to explain to the editor. Other writers, however, dazed by the editing or pressed for time, return the manuscript without a single change.

I recommend a different course. Examine the editing carefully. Imagine the editor sitting beside you asking questions such as, "What about saying it this way?" or "Did I understand this correctly?" Then make changes selectively.

Thoughtful writers improve an edited manuscript in two ways. First, they catch mistakes—perhaps a word left out when the editor combined two sentences or a verb that no longer agrees with its subject. Second, they interpret what the editor is doing and offer further assistance. For instance, in editing a manuscript about India I cut several paragraphs and then struggled to write a transition connecting the two loose ends. I came up with only one uninspired sentence. The author spotted the problem and supplied an anecdote from his travels that made the transition more coherent and added some engaging information. In other cases I have cautiously trimmed rambling introductions, and the perceptive authors have noticed and then suggested even more drastic pruning. But wise writers don't go overboard. They are conservative about inserting changes in an edited manuscript—sometimes making only one every page or two.

Unfortunately editors can't explain every change because they may make twenty or more on each page. What's more, they can't convey in the margins of your work all the intricacies of English they have learned over the years, so you have to decipher most of their revisions on your own. But if a few puzzle you, ask about them. And remember that when editors read your prose, they can't always understand what you have in mind, so they may occasionally twist the meaning in the editing. When this happens, I'm sure many authors curse the editor for being so stupid. But if you are clever, you will check to see if your original wording is ambiguous.

People know that editors like to ferret out excess words and phrases, although I'm not sure they all understand why. Yes, paper is expensive and publishing budgets are tight, but that isn't the main reason I trim excesses. When I'm editing, I often visualize myself hacking through a jungle with a machete. Writers set out a path for readers to follow through their research, but by including too many additional words, facts, diversions, or contradictions, they obscure the route, and readers may lose the way. As an editor, I chop away underbrush so the trail is absolutely clear. If the theme of the article is valuable, I want every reader to grasp it and not be distracted by unnecessary verbiage.

Good editors are not capricious. They don't acquire their own peculiar styles, although many authors are convinced that they do. Competent editors know how language works and want every article to be comprehensible to the broadest possible audience. Editors rely on standard reference books, particularly the *Chicago Manual of Style*. True, not every editor emphasizes the same things; each has strengths and weaknesses. I, for instance, seem to be one of the few people in the Western world who understand ellipses, but on commas I'm often vague. Sometimes editors simply have different ways of solving the same problem. Authors confronted with differing suggestions from two editors may mutter about the quirks of that breed, yet serious writers have discovered that each editor can teach them something new.

Occasionally authors protest that editors obliterate their personal writing styles. I'm skeptical, and I don't think this complaint is appropriate in the academic world. Idiosyncratic language may succeed in fiction, but I contend it is out of place in nearly all scholarly work. In journals we need to present research simply and clearly. Unlike novelists and poets who may play with words for evocative effects, scholars must handle them with precision. Exotic sentence structures, odd punctuation, and words used in unfamiliar ways bewilder readers trying to grasp the meaning. Editors who know the nuances of English can direct writers to the standard ways it is written and understood. Using accepted English, however, never prevents writers from producing lively, vigorous prose with a personal signature.

I suspect that the toughest part of publishing for most authors is the waiting. The long stretches between letters from the editor and the vast expanse of time between acceptance of a manuscript and publication are frustrating. I know—I'm an author too. I have no solution except to warn authors that this is normal. Editors are usually conscientious about minimizing delays, but they can't produce publications instantly. Most editors are juggling several journal issues at once, with perhaps three dozen manuscripts or more. In addition, they are coping with budgets, advertising, new printing technology, and a host of other problems authors can't imagine.

One last note. If you think an editor has improved your article, thank that magician.

CULTIVATING AN AUDIENCE

When your article finally bursts into print, there is one more step academic authors often neglect. After pouring time and effort into research and writing, scholars are often disappointed by how little reaction they receive. But publishing an article is not enough to launch it into the scholarly world today, particularly if the journal is specialized. However, there is a remedy: distributing reprints.

In the past, scholars shared only a handful of journals in their field, and nearly everyone subscribed to them. If your article appeared in one, you could be sure colleagues would see it. But today dozens of new journals enter the market every year, and even major university libraries don't carry all the significant ones. Scholars are overwhelmed with material to read, so simply steering your article into print doesn't guarantee that your colleagues will find it.

As an editor, I have watched a few brilliant articles fade quietly into obscurity, while some fairly ordinary pieces catapulted their authors into symposiums and new research projects. The difference was that the savvy writers handed out reprints. I have also surveyed authors by mail asking about reaction to their articles, and I discovered that nearly all the response they received in the first year or two—such as letters commenting on their work and invitations to deliver papers—came from reprints they had mailed.

Despite the benefits, some scholars feel uncomfortable passing out their work. To them it smacks of unattractive self-aggrandizement. But I see it differently. We work in a community of scholars linked by the flow of ideas. Within this community, specialized publishing has succeeded almost too well, overloading the lines of communication. Many thoughtful authors, excited by their research and eager for discussion, address the problem head-on by sending reprints to colleagues. Distributing reprints broadens the arena of discussion and speeds up the reaction time—two results that scholars should applaud. Of course, you can overdo it. Nobody likes a blatant self-promoter.

I suggest authors circulate a minimum of twenty reprints, and save some copies for job applications, promotion decisions, and grant applications. As your intellectual network expands, aim for a mailing list of fifty to one hundred people for major articles. The heart of your list will be scholars working in the same specialty (whether they reside in geography or other disciplines) but include colleagues with other research interests, because they should be aware of your current work when talking to students and administrators. Supply copies to everyone who participated in your research: colleagues who read drafts, city employees who supplied information, citizens who responded to interviews, or curators who found photographs. You can also use reprints to get acquainted with other scholars; many will respond with letters and articles of their own. In composing your mailing list, be imaginative, because some articles would be welcomed by politicians, news reporters, environmental groups, planning officials, or radio interviewers.

Once colleagues see your article, they may assign it in classes, cite it in their own work, ask you to give a presentation, include you in a research project, or simply be prepared to talk with you at a meeting. Even colleagues who merely glance at the introduction may pass on valuable research leads to you. But none of these things will happen unless people know about your work.

TAKING STOCK

I'd love to say that once you understand publishing, writing articles is easy. But it's not true. For most of us, writing is hard work. Nevertheless, understanding the publishing process, from choosing a journal to sending out reprints, has other benefits. Authors who learn how journals operate certainly improve their chances of having a manuscript accepted. Knowledgeable writers also benefit more from their encounters with editors. And when scholars have the practical aspects of writing and publishing under control, they can devote their attention to the most critical task in developing a strong research article—thinking.

Barzun, J., and Graff, H. F. 1970. *The Modern Researcher*. Rev. ed. New York: Harcourt Brace Jovanovich.

Bernstein, T. M. 1965. *The Careful Writer: A Modern Guide to English Usage*. New York: Atheneum.

Brunn, S. D. 1988. The manuscript review process and advice to prospective authors. *Professional Geographer* 40(1):8–14.

Chicago Manual of Style. 1982. 13th ed., rev. and expanded. Chicago: University of Chicago Press.

Directory of Publishing Opportunities in Journals and Periodicals. 1981. 5th ed. Chicago: Marquis Academic Media.

Graves, R., and Hodge, A. 1979. *The Reader over Your Shoulder*. 2d ed. New York: Random House, Vintage.

Hanson, S. 1988. Soaring. *Professional Geographer* 40(1):4–7.

Hart, J. F. 1976. Ruminations of a dyspeptic ex-editor. *Professional Geographer* 28 (3):225–32.

Lee, D., and Evans, A. 1984. American geographers' rankings of American geography journals. *Professional Geographer* 36 (3):292–300.

Robl, E. H., ed. 1983. *Picture Sources Four*. New York: Special Libraries Association.

Strunk, W., Jr., and White, E. B. 1989. *The Elements of Style*. 3d ed., rev. and expanded. New York: Harper & Row.

Turner, B. L., II. 1988. Whether to publish in geography journals. *Professional Geographer* 40 (1):15–18.

Tips and Comments on Compiling a Multi-Authored Collection

Martin S. Kenzer

Compiling and editing a multi-authored collection of papers is unlike writing a book or a textbook; it differs entirely from writing an article; it is significantly distinct from editing a journal; and it is different from putting together an anthology of previously published papers. In some respects, it can be considered among the most frustrating academic endeavors a scholar can undertake; yet, it is also a highly worthwhile and satisfying experience when the task is (finally!) complete. But university administrators and tenure committees rarely regard such volumes as equal in importance to single-authored (or even dual-authored) books. Those inclined to undertake such a project need to realize at the outset—long before they begin—the necessary time frame and the exasperations involved. For the credit they will receive, their time may be better spent writing their own book, or even writing two or three well-placed journal articles.

There are no simple guidelines to follow when editing multi-authored collections, but it still need not become a frustrating or hair-raising exercise. While I cannot anticipate every potential compiler's exact circumstances, in this chapter I will address the most frequently encountered problems in compiling and editing a multi-authored collection. I will draw primarily on my own editing experience, but also on the comments of others who have brought such volumes to fruition.

BEWARE: IT'S *YOUR* PROJECT

The first and foremost thing to keep in mind when considering editing a multi-authored volume is that the papers are all commissioned—i.e., *you* request them. The manuscripts are not written independently and submitted, as is the case with journal articles. In other words, the motivating force behind the papers comes from *you,* not from the respective contributors. This obvious but important fact has several implications. You should not, for instance, expect contributors to show the same degree of sustained interest in the project as they might for their own projects. Similarly, multi-authored collections do not just happen: a group of like-minded academics rarely, if ever, approaches you with finished papers, saying, "We have these completed papers that just happen to be on the same topic; why don't you go through them, edit where necessary, and find a press that will publish them." Instead, *you* must first initiate a worthwhile project; *you* must then approach a good number of potential authors to see whether some (or any) of them are interested in your proposed volume; and *you* must make sure that the final product you envision is also what your potential contributors have in mind. After you have convinced an adequate number to undertake their respective chapters—bearing in mind that some will never start, much less finish, their papers, which means you will need to solicit back-up papers as well—then, and only then, *you* must approach editors and convince them that the highly motivated crew of authors you have assembled will, in fact, deliver. "No problem," you say? Unfortunately, that is only the beginning.

The world of academic scholarship—notably conferences, reports, books, journals, and textbooks—functions according to a series of deadlines. Do not, however, expect to adhere to deadlines when compiling a multi-authored volume; you must establish them, of course, but do not, under any circumstances, rely on them. You and the press publishing the collection will take deadlines seriously; many of your contributors unfortunately will not. They will have prior and intervening commitments, which will most often take precedence over their contributions to your volume. For you, the volume editor, the finished product is an important undertaking; for contributors, however, it is more often than not merely another obligation, which may demand that they drop more important projects to meet your expectations.[1] Remember that it is *your* project and *your* enthusiasm, and in most cases your contributors will not share that enthusiasm to the same extent. They will react positively at the outset; they will occasionally give you support and feedback during the project. But they cannot be expected to carry your excitement for the finished product to the project's completion. In fact, your own energy will wane from time to time. Nonetheless, you should set deadlines. Without them you will *never* complete the project; with deadlines you will complete it late.

Bear in mind that even a few who respond favorably to your initial query about contributing an essay will in time fade from the project. Some will ask to withdraw due to a variety of excuses; others will promise you week after week—month after month in some cases—that the manuscript is almost complete, and yet they will never produce it. Never forget the old expression, "The check's in the mail." You will hear the academic equivalent many times over. It is a sad state of affairs, but only a minority of professional geographers are prompt or conscientious about meeting deadlines. Many never return phone calls, and fewer still answer their mail. So, you should not expect them to stick to a deadline. Once you realize this, your expectations will change accordingly, and your level of frustration will subside proportionately.

[1]I am not suggesting that all contributors are delinquent in fulfilling such commitments; some are extremely prompt and always abide by deadlines. Nor am I insinuating that others don't have perfectly valid reasons—sometimes beyond their control—to miss deadlines. I have been at both ends of this process—volume editor and contributor to someone else's book—and I, too, have missed deadlines. I appreciate that circumstances often prevent even the best-intentioned writers from meeting obligations. But the very nature of multi-authored collections means that deadlines are generally less important to the individual contributors than they are to the volume editor.

On the other hand, one positive aspect of putting together an edited collection is that you will learn whom to ask to contribute to subsequent projects. In a sense, it is a way to discover those with whom you can work and whom you can trust to deliver. It takes at least one editing endeavor to learn who the deadline-conscious individuals are, but you will both appreciate those people and come to rely on them for future writing commitments. They are the backbone of the discipline.

When the manuscripts do arrive, do not expect them all to be precious gems. Such expectations are unrealistic, even from the best of authors. But, beyond this, be aware that the articles you read in journals are readable, for the most part, because editors spend untold hours turning mere words into well-crafted prose. Moreover, the contributions will rarely break new intellectual ground. The collection will almost surely be what a polite reviewer will consider "uneven"—meaning that some are quite good, while some are assuredly weak. A number of authors take this sort of endeavor less seriously than the writing of journal articles or their own books, so the quality of contributions can vary considerably.

WHAT YOU WANT IS NOT WHAT YOU'RE LIKELY TO GET

The uneven quality of contributions can pose a problem for editors of multi-authored collections: what to do about "garbage"? Remember, you commissioned the essays, so you must then decide which are worthy of inclusion, which are okay but need reworking (to various degrees), and which are simply unsuited to the volume. You cannot and should not sacrifice quality simply because you asked a colleague—and possibly also a friend—to write something for your book. So, don't kid yourself into thinking that you will want to include all the completed contributions. You may have to work closely with some authors, helping them struggle through several drafts of their manuscript, before you receive an acceptable chapter.

Furthermore, a dilemma arises at this point. You asked, and possibly begged, these individuals to contribute to your collection; generally, they did not approach you. Unlike the case of a manuscript submitted to a journal editor, frequently only you will see the finished papers before they go to press. A publisher may ask for outside reviews, but often this is done on the basis of an outline proposal, and publication decisions are based on the merits of the anticipated collection as a unit, not on individual chapters. The publisher will assume that all contributions are scholarly. Thus, you are often the only one responsible for deciding whether an essay should be included or not. Moreover, even if you did not know the individual authors before, you will probably have developed some sort of friendship with them through correspondence or telephone conversations. It won't be easy to tell one of them that an *invited* paper is now unacceptable for the collection—especially after the author has met your deadlines and has *finally* finished it. The bottom line, however, is that a paper that is garbage—unless you personally are willing to rewrite it, which is more common than you might anticipate in such collections—will do your book no good and should be dismissed as unworthy of publication.

The best thing to do is to establish criteria for what is acceptable and inform all potential contributors beforehand that unacceptable submissions simply will not be included. Of course, if you are working with experienced authors who have produced for you before, this is unnecessary. *Always give yourself the option to reject a manuscript*—but make this fact clear to the contributors before they write a single word. No one wants to labor over an invited manuscript that will later be turned down without warning. Many people assume that, if they are asked to contribute to an edited collection, the editor will accept whatever they submit, regardless of quality.

A related concern involves contributors who send you papers that no longer have any bearing on the topic on which they were asked to write. This, too, happens more often than one might think. It can result from a lack of communication between editor and author; it can also occur because some authors are poorly organized and tend to wander in their writings, thus losing track of what they first started to write about. Finally, it can happen because some people see an invited paper as a forum for personal ruminations, rather than as an opportunity to contribute to a well-integrated set of essays with a specified theme. *Do not allow this to happen*. If you request a paper on the diffusion of dairy farms in the Upper Midwest, but you receive a manuscript on the distribution of tractors in Arkansas, send it back to the author. Explain that this was not the topic mutually agreed upon and that it simply cannot fit in with the other chapters.

One way to avoid this, in part, is to request a general outline from the prospective authors beforehand. An outline can serve several critical functions:

1. You learn whether or not you and a potential contributor are thinking along the same lines.
2. It gives you an early opportunity to alter the focus of contributed papers and to offer suggestions on method and organization.
3. It provides you with something concrete from each author that will help convince a press to publish your book.
4. It gives you something in writing to refer back to when an author ultimately produces something quite different from the agreed-upon outline.

If authors cannot produce a coherent outline, you cannot expect them to deliver a quality manuscript.

Sometimes you will receive completed manuscripts that may not be precisely what you had in mind but are nevertheless scholarly and deserve to be published. Together, several such papers might provoke discussion or perhaps stir you to reconsider your own thinking on the topic. In fact, the finished products may redirect your original scheme for the volume. In this case, if you decide to accept and include such manuscripts, you are then obligated to change the initial organization of the book. It may be that you should reexamine your intended chapter outline. It also may mean that when you incorporate the papers into the volume, the result may be an uneven balance. Balance is not a prime concern, however—quality is. *You* solicited the chapters, *you* agreed to accept what was written, and it is now *your* responsibility to change the arrangement of the volume to match the papers at hand. Finally, as editor you are obligated to tie the collection together and give it focus in your introduction. If the final chapters don't quite resemble what you envisioned at the outset of the project, this can be rather difficult. However, never forget that this is your project, so it is your obligation to pull it all together in the end and to make it a coherent whole.

FURTHER HEADACHES

When editing a multi-authored collection, be aware that you will have to accommodate the idiosyncrasies of perhaps twenty or more academics, all of whom believe that their way is correct. No matter how you try to avoid it, some will create unneeded problems. Some think that their words are jewels, so they will refuse to let you touch their prized sentences. Others will send you a 3,500-word paper when they were clearly asked to produce one of 7,500 words; typically, however, it is the other way around. Some will say, "But I've been publishing papers for twenty years, and I've always done it this way." Some, no matter how many times you tell them, will refuse to follow directions with respect to style or other specifications. There are a lot of "independent thinkers" in geography.

There is little you can do about this, except to provide a set of guidelines, and the sooner the better. If nothing else, this will give you something on which to base your rejection when authors refuse to comply with your instructions. The reference section of a manuscript is a particularly common item for authors to misconstrue. No one likes to redo references to conform with a particular journal style, and many simply will not do it for an invited collection. When they do, some merely hand the instructions to a secretary and say, "Put it in this style!" If you are fussy about such things, explain this at the outset, and let the contributors know that a manuscript that arrives in a style totally unlike the one detailed in your guidelines will not be accepted. Leave your options open. The bottom line is that you probably will not get what you expect, so don't expect it, and, by all means, don't anticipate perfection.

You will sadly learn that some academics *expect* you to rewrite their papers for them. A number of geographers, some well-respected in the field, will submit what can only be considered trash, anticipating that *you* will do what they were unable to do in the first place. For instance, you might receive a manuscript that has the germ of an idea buried in it but is so poorly organized and so awkwardly written that there is absolutely no hope that a reader will unearth that idea in the paper's present form. So, you will invest hours in editing it. When you return the completely rewritten paper to the author, expecting him or her to refuse point-blank to make so many changes, instead you will quietly receive a clean, recomposed copy that incorporates virtually all your suggested changes, without a whimper of protest from the author! Was this expected from the outset? Did the author assume that you would turn an eyesore into a masterpiece? You will never know for sure, but it was probably taken for granted that you would do the dirty work. Be prepared to rewrite some contributions; also, be prepared to encounter the other extreme, contributors who will scream at every change, no matter how minor. Editing a multi-authored collection sometimes seems like a no-win situation.

I briefly mentioned the potential problem of asking friends to contribute to the project. Friends, perhaps more so than others, often assume that you will accept anything they send you. Occasionally, they may infer that you are asking them to contribute because you cannot find anyone else to write a particular chapter, and thus they may expect even greater leeway in what they write. Soliciting friends to become part of a multi-authored volume is frequently asking for trouble, and sometimes it results in the loss of a friend. Be judicious, and think twice before extending the invitation.

Similarly, I recommend that you do not ask close friends to coedit a multi-authored collection, unless there is no other option. Whenever more than one person undertakes a project, specific terms of the division of labor need to be arranged beforehand, but this is even more critical in a coedited collection, especially when the editors are friends. If your coeditor is at a different university, possibly at the other end of the country, decision making will be a slow and cumbersome process. Do not ruin a good friendship. Instead, if you want assistance, ask a colleague or a professional editor you respect to help you coedit—not a friend.

DETAILS, DETAILS, AND MORE DETAILS

A note on reading proofs and authorizing manuscript changes is in order here. Unlike a journal editor or the sole author of a book, you are frequently placed in a somewhat awkward position as editor of a multi-authored collection. Consider the following scenario. You have solicited, received, and edited twenty manuscripts for your book. This means that you and each author have come to an agreement on the style, punctuation, content, and length of the chapters. It has taken, say, eight months—a very optimistic estimate—to arrive at this stage. You then send the twenty chapters to the press, only to have their copy editor return the manuscripts reedited, often significantly, to suit house style! Now what? Do you send the manuscripts back to the authors and ask for their approval? This not only may take a great deal of time, but it may cause further frustration on the part of authors who thought that the editing process was over. Or, do you decide not to return the manuscripts and instead approve the new editorial changes yourself? This, too, is certain to create problems when the chapters appear in print and they differ from what you and each author agreed upon. Likewise, when the press sends you proofs, do you mail chapters to their authors to examine—knowing full well that some will let the proofs sit for days, even weeks, before they attend to them, which may mean that a project already behind schedule will be even later—or do you go over the proofs yourself, knowing in advance that you will get the job done sooner, but you will be responsible for every typographical error that appears in the book?

There are no easy solutions to these problems, but you need to anticipate them at the outset. If you are dealing with an understanding editor at the press, or one who trusts you implicitly, perhaps because of having worked with you before, then one way around these issues is to agree on them in advance. For example, you and the editor may agree that you will be the "final" editor—i.e., no copy editor will alter what is sent to the press. Or, you may agree that you will be a compiler, and not the copy editor. In the latter case, you won't copy edit contributions; instead, you will simply organize the project, helping the authors through their assigned chapters, and then rely on the copy editor at the press to do all the dirty work. There are obvious advantages and disadvantages to either approach. What is important is to put the project into proper perspective, remembering what has already been written above about motivation and individual idiosyncrasies.

Multi-authored writing projects lend themselves to problems that rarely arise in single-authored or coauthored books. Contracts, for example, can be particularly troublesome. When you sign a book contract with a press, you will agree to certain conditions. Royalties, for instance, are generally a part of every contract. In a multi-authored volume, however, royalty clauses can be a terribly sticky issue: do you include each author in the royalty clause? Often the press will have an established procedure to deal with this, but it is typically not press policy to pay each author a royalty; as you can well imagine, the headaches involved in figuring royalties are multiplied many times over for books with twenty or more authors. Further, as volume editor, you will do most of the work to bring the project to fruition, so royalty payments should go to you. In most cases, moreover, scholarly geography monographs do not sell enough copies to entail worry about such matters. Most contributors realize this, but some will still fuss about royalties.

If you are editing a multi-authored textbook, where there is significant sales potential to be realized, then it is only fair to pay each author a royalty in the form of a one-time honorarium. In such cases you, too, will receive royalties from the book, but only *after* the one-time honoraria have been paid. In other words, the authors' one-time royalty payments come "off the top" of the book's sales, and you will then receive whatever is left over. Be sure both you and your contributors understand this. There are no hard and fast rules to follow; circumstances will dictate the amount of honoraria. One way to satisfy your contributors—and remember that an edited, multi-authored volume would never come to pass without each and every contribution—is to assure authors that they will receive a number of free copies of the finished product. A press will be much more willing to agree to this arrangement, since copies of a book are relatively inexpensive. If your book will appear in both a clothbound and a paperbound version, make sure the contributors receive copies of each.

If your anticipated volume will have a significant amount of artwork or graphic materials, remember that permission has to be obtained for each piece, unless it is an original contribution. It is the individual author's responsibility to request permission, but if they don't do it, you will have to. You will be stuck in the middle again, between the press and the authors, so you will hear complaints from both. Make this responsibility for obtaining permissions clear to your contributors at the outset. Large presses frequently have standard forms for requesting permissions, so ask your press editor for them.

Whenever you deal with a press, there is always the possibility that the editor with whom you've been working will no longer be at the press when it comes time to submit your completed manuscript. Scholarly, trade, and textbook presses alike, for a number of reasons, have a significant employee turnover rate, and this can pose unforeseen problems. It can be frustrating to work with a press editor on your own book, only to have that person replaced by someone else who has to begin anew on your project. When you're editing a multi-authored collection, however, this sort of personnel turnover can result in even more problems. One thing to keep in mind, whether you are the sole author or an editor responsible for a number of other authors, is to *get every last detail in writing!* Contracts are always written. But often agreements and changes on peripheral matters are made over the phone, and although they may seem minor at the time, they can come back to haunt you. You may pass such agreements on to your contributors, only to learn at a later date that your press editor is no longer employed by that company, and no one else knew of your verbal agreement. Get it in writing.

THE CHICKEN OR THE EGG?

I mentioned above that you should solicit manuscripts first, and then contact a press to publish the papers. This is not the only way to approach a multi-authored collection. As a volume editor, you are caught in a precarious position: major presses frequently won't agree to publish a collection until they see the finished product—remember that they are well aware of the pitfalls of such projects, many of which are aborted in midstream and never come to pass—and, conversely, many authors won't agree to pen a chapter for a project where no contract yet exists. What to do? Basically, you have a sales job ahead of you. You either have to sell potential contributors on an idea (and to convince them that you can get that idea into print in a relatively short period of time), or you must sell a press on the quality of your potential contributors (and their ability to produce something of value in a reasonable period of time). Both tactics require certain skills.

If you've dealt with the press before, or if the press is sold on the idea, then it may be easy to secure a contract. Likewise, if you are dealing with authors who have contributed to prior projects you've edited, or if they know your track record and your ability to obtain a contract, then it may be relatively simple to get them to write a chapter. But, if you are starting out on your first such project, I strongly recommend that you line up your authors first—and don't forget backup authors—and then approach a press. Often a good idea and book outline, along with abstracts and/or chapter outlines from each contributor, will sell the press on your proposed volume.

To approach a press without a pool of authors, no matter how good your idea may be, has obvious problems, even if the press offers you a contract immediately. For example, what happens if you have obtained a contract but cannot find authors to do the chapters? You could always write some of the chapters yourself, but if you wind up penning several of them, you might as well write the volume yourself—in which case you will receive far greater credit. Also, to obtain a contract for a multi-authored book where no contributors exist is something like selling a collage before it has been created. You have no idea who the final contributors will be or what they will agree ultimately to write on, and the press might not be willing to bend as much as you will when your authors redefine the volume's theme. If you have a tremendous idea for a collection for which you know any press will be eager to sign a contract, no matter who the authors turn out to be and irrespective of the final focus and quality of each chapter, get the contract first, and then seek out the lucky authors. Under all other circumstances, however, I suggest that you do it the other way around. At minimum, if the manuscripts are complete but a book contract still appears unlikely, you can then consider guest editing a special issue of an appropriate journal as an outlet for the papers.

One way to test whether or not you have a publishable idea—i.e., to determine whether you will be able to secure enough potential authors even to consider editing a collection—is to solicit papers for a special session at a local or national meeting. If you are unable to organize a session on a given topic with only three or four speakers, then either you are not a good organizer—which automatically means you should not attempt to organize a multi-authored volume—or there may not be enough interest in the topic to warrant pursuing the matter further. However, if you do organize a session and receive inquiries from other interested potential contributors—word travels fast in academic circles—or if you find "standing room only" at your special session, you probably have a worthwhile idea that can be augmented and turned into a published volume.

To organize special sessions at meetings has yet another advantage: it goes a long way toward assuring you that at least some of the authors can produce, and that their papers, if no others, will be completed. Papers read at meetings necessarily will need to be altered for publication, but this is a major step in the right direction. Once the presented papers have been completed, then others may opt to join your project because they are confident that their work won't be for naught.

Assembling a group of potential contributors, whether at a professional meeting or elsewhere, serves a third purpose: it provides an occasion for the authors to meet one another. Because the chapters of multi-authored collections have a related focus, readers might assume that the authors already know each other, and communicate on a regular basis because of their kindred interests. Not so. Many academics work alone, often in isolation from one another, and frequently they have little direct contact with others who are doing research on the same general topic. Further, not all multi-authored collections involve contributors who conduct research in similar areas. This volume, for example, draws on geographers with divergent backgrounds, none of whom work in the same subfield. I organized the prospective contributors independently of a meeting, but I later took advantage of their attendance at the Phoenix, Arizona, meeting of the Association of American Geographers (1988) and asked them to meet for breakfast. So, if you are planning to edit a collection of papers, I suggest you first organize a thematic session at a professional conference, followed by (or in conjunction with) an informal meeting that includes the session participants and anyone else you would like to have write a chapter for your book. This has the added dimension of personalizing the project and can foster motivation through a sense of group participation. Sometimes merely knowing that endeavors are part of a group effort can help to stimulate creativity and productivity.

SO. . .IS IT ALL WORTHWHILE?

Given the primarily negative comments made above, you would be fully justified to ask a number of important questions at this point. For instance, considering all the inherent problems I've noted in producing such collections, why should someone invest the time necessary to complete a multi-authored book? Why not, for example, merely write all the chapters yourself? After all, from the standpoint of peer respect and/or tenure and promotion considerations, a sole-authored or even a coauthored volume carries more weight than do edited collections such as this one. So, why burden yourself and go through it all? Obviously I believe the endeavor *is* worth the effort; if not, this volume never would have come to pass.

First, we all have a limited amount of time to invest in any given project. In other words, despite the drawbacks noted in organizing and bringing such collections to fruition, few of us have the time to write many full-length books. Instead, we organize such projects and ask others to write the bulk of the chapters. Thus, we endure the hassles of editing a volume so that the projects that seem most essential will be completed. If we opted instead to devote additional time to such projects and write the books ourselves, by necessity this would subtract from the time required to pen articles and single-authored books in our areas of specialization—something many of us believe is more important to our personal research agendas.

Second, it would be foolish and presumptuous to contend that any one person knows enough about all the topics in a volume to undertake the writing alone. Certainly a multitude of authors will bring a broader and more diversified set of questions to bear on a given subject than any single author possibly could. In this book, for example, the wisdom and strengths of specialists in their respective areas of expertise were solicited, to impart insights that no one geographer alone could have provided effectively. Hence, the readers benefit exponentially because they learn something from each author's perspective, rather than reading a single author's point of view throughout.

Finally, there are nonquantifiable aspects to multi-authored volumes that are difficult to articulate. There is an almost electric quality about a group effort that seems to be lacking in single-authored projects. In a sense, group efforts take on a life of their own, which seldom happens when authors work independently. Likewise, there is a special feeling in organizing a group project. It's as though the end product will somehow be better because enough people took valuable time out of their busy schedules to contribute to the effort—*your* effort. The rewards in academia are often few, but one of them is the knowledge that you were able to generate an idea that a number of your colleagues considered worthy enough to lend their time to. And, when the enterprise finally comes to pass, when all the difficulties have been surmounted and the problems are behind you, there is an added touch of exhilaration in knowing that you've succeeded in beating the odds.

How to Win Extramural Research Funds

Ronald F. Abler

Ask, and it shall be given you. . .
For every one that asketh, receiveth.
Matthew 7, 7–8

If you need external funds for research, this chapter will tell you how to obtain them. Despite the biblical injunction, most people find it difficult to ask for money. Why one asker receiveth and another receiveth not is often a mystery. I hope to allay your fear of asking and to demystify proposal preparation and evaluation by explaining how to formulate a fundable research problem, how to select a funding agency, how to craft a strong proposal, and how to interpret responses to your proposal.

DO YOU NEED EXTERNAL FUNDING?

There are good and bad reasons for seeking external funds. Good reasons are that your research is compelling and that it cannot be conducted without external support. Bad reasons are that you want additional salary or that your department chairperson wants you to capture external funds. Before embarking on a quest for funds, examine your motives and needs carefully.

Unless you are exceedingly clever, winning an award will entail toil and trauma. A chance to satisfy your driving curiosity will sustain you through the weeks or months of work needed to prepare a competitive proposal. Other motivations will not. External research is not a money-making proposition; you will probably spend more time preparing proposals than you will win in extra salary. Proposals written to mollify chairpersons or deans waste everyone's time. They convey little excitement or commitment and are quickly dismissed by funding agencies. In short, if you need money for your research, read on. If your research does not *demand* external funds, don't waste your time applying for them.

If you need research money, don't be shy about asking for it. Funding agencies are designed to give money. They need requests for funding to stay in business, much as doctors need patients. Program officers are judged by the results obtained by scientists whose research they fund. The more requests program officers receive, the more chances they have to select projects that will yield worthwhile results. Numerous requests also give a program officer a stronger claim on agency funds. By the nature of their jobs and goals, program officers and funding agencies want proposals that address important research problems, and the more the better.

FORMULATING A FUNDABLE RESEARCH PROBLEM

A research problem is a sharply defined question that you can answer. When you have formulated a research problem you can frame your question (hypothesis) in a single sentence, specify facts that constitute evidence for the possible answers to the question, describe in detail the analytical techniques needed to process those data, and specify how you will know whether you have answered the question after you have analyzed your data.

A research topic is not a research problem. Understanding the relationship between fossil fuel consumption and atmospheric carbon dioxide, for example, is an important research topic. But without a specific question, measurable variables, and a logical argument connecting the measurements to the question, it remains a topic. Many purported research proposals consist solely of general research topics. They never come to grips with what the applicant intends to contribute in the way of specific answers to specific questions.

A research problem is fundable if the prospective answer excites others besides you to the degree that they will give you money to obtain that answer. A perfectly formulated research problem may promise results that interest nobody but you. Therefore, you must be able to demonstrate why the answer to your question is relevant and important to those to whom you will address your request for funds.

SELECTING A SOURCE OF EXTERNAL FUNDS

You can't receive unless you ask. You won't receive unless you ask at the right place. Once you have formulated a tractable problem, your next task is to find the most promising source of funds for your problem.

Identifying Organization Biases

Every funding organization supports some topics and ignores others. Most funding agencies issue statements describing their goals. Read them carefully, and act accordingly. Lists of projects previously funded are especially helpful, and directories can provide some guidance. Your university research office, your colleagues, and your scholarly society may provide advice. But determining which of the thousands of funding sources in North America is best for your proposal is your job. You know what you want to do, so you are best qualified to evaluate interest on the part of an agency or program officer.

The National Science Foundation (NSF), for example, has two biases no applicant should ignore. First, it supports basic science. Second, NSF is reactive; it responds to investigator-initiated proposals.

The boundaries among basic science, policy research, and applied science are fuzzy, and most NSF program officers would be hard put to provide a precise definition of "basic science." But they and the reviewers they employ recognize its presence or absence in a proposal. If your research is grounded in theory and could change the way geographers think about the world, NSF will be receptive to your proposal. If your project's primary justification is better policies or more effective applications of known principles, you are unlikely to succeed at NSF regardless of your project's importance or urgency.

The foundation rarely issues requests for proposals (RFPs). You must decide what you want to do, and then convince reviewers that your project is worthy of NSF support. Never ask NSF program officers what they are interested in funding. That's a silly question, even though underemployed assistant deans and assistant vice presidents for research wander NSF halls asking it. Asking such a question betrays inadequate homework; it is sure to remind a program officer of an urgent appointment elsewhere.

Note well that a question that is a gaffe at NSF may be the right question to ask elsewhere, and vice versa. Some funding agencies and institutions do not entertain unsolicited proposals. Instead, they describe the research they wish to fund and accept bids for specific projects. When dealing with such an agency, asking about general or specific preferences is appropriate. If you want to conduct research supported only by such an agency, you must convince those in charge of research funding to issue an RFP for your project. The process of doing so is much the same as described below, except that it is done through informal discussions rather than by a formal proposal.

Regardless of agency biases, you should first decide what you want to do and then identify the funding source that is interested in your problem. You will earn little respect and have little success if you are willing to "do anything." Personal commitment to a problem guarantees results. If you are willing to do anything, you are committed to nothing.

Preproposal Contact

Assuming you have narrowed your search to agencies whose biases match your needs, your next step is to contact their program officers. If the contact is face-to-face or by telephone, be prepared to explain—in five minutes or less—precisely what you want to do, why it is worth doing, and approximately how much money you need. If the contact is by mail, explain the same items in no more than three pages of text. If you cannot summarize your research problem in five minutes or three pages, you are not ready to seek research support.

Unless you know the program officer is well-versed in your specialty, your prospectus should be intelligible to a generalist. Geography is a diverse discipline, and no program officer can be conversant with all its facets. In many agencies and foundations, you will be addressing nongeographers. Make your case in terms program officers are sure to understand.

Be sensitive to program directors' reactions to your research description. (If the interview is oral, prepare a list of questions and take written notes of responses.) They may state flatly that their agency is or is not interested in receiving a formal proposal, but the message is often more ambiguous. A statement that your project is appropriate for review by the program or agency may be as much encouragement as you will receive from a seasoned program officer. If you receive more than that, so much the better. An observation that the program has not funded such work in the past (or for a long time) is a roundabout way of saying that submitting your project to the program would be a waste of time.

If you detect discouragement, respond as follows: "It sounds as if your program might not be the best place for my idea. Can you suggest a program or agency that funds the kind of work I want to do?" That sally will give the program officer a chance to correct any misinterpretation of the response, and it may also yield some alternative sources of funding.

If you detect encouragement, respond in this way: "If I interpret what you've said correctly, a strong proposal along the lines I've suggested would have a reasonable chance of being funded at your agency. Based on what I've told you, can you suggest ways my proposal could be strengthened, or do you note weaknesses I should eliminate in my formal submission?" That will give the program director an opportunity to correct misinterpretations on your part and also to identify points or topics that should be emphasized or avoided.

Make sure you understand when and how your proposal should be submitted. If written guidelines for proposal preparation are available, make sure you have an up-to-date version as well as current copies of any required forms. Finally, if you have not yet seen one (you should have!), ask for the organization's most recent award list. Past funding patterns are not perfect predictors of future awards, but they are better than anything else.

Preproposal Modifications

Think long and hard about any suggestions the program officer has offered, and incorporate what you can. But do not compromise the project's integrity. Agency biases and interests are usually flexible. You will do better with a strong project that does not exactly meet an agency's biases than you will with a project that has been weakened to meet them.

Next, calculate a budget for your project, using the format required by the agency or agencies to which you will address your request. Do it their way, allocating costs as each agency requires. Now compare your project and budget with those listed on the most recent award list available. If your topic and budget fall within the ranges of subjects and amounts the agency has been funding, you are most likely on the right track. If your topic or budget deviates significantly from recent awards, check to make sure you have not gone off on a tangent. Make sure that the program officer is aware of the deviations you have identified, and that they cause no concern. Use this final preproposal conversation to resolve any questions about what kinds of costs the program will support, the allowable length of projects, and so forth.

Simultaneous Submissions

If you have identified two or more funding sources that are interested in your problem, make sure you understand each source's attitude toward submitting the same or similar proposals to other agencies. Most agencies encourage simultaneous submissions. They realize that a good project may excite several sources and that an applicant may be able to create a complete package with partial funding from several agencics. But all agencies will want to know about your other applications. Be sure you report simultaneous submissions completely and in the manner required by each agency.

Be prepared to tailor your proposal and budget to each source to which you apply. Proposals to fund the same work submitted to NSF, the National Geographic Society, and the National Endowment for the Humanities will require different narratives emphasizing each agency's interests, different budget formats, and different submission dates.

RESEARCH PROPOSALS

A research proposal consists of an answerable, generalizable question, a plan for answering the question, and a research budget. Each agency or foundation wants research proposals submitted in its own format. Follow their instructions scrupulously.

The Question

A research proposal that begins, "The question I wish to answer in this project is. . ." is well begun. As reviewers read on, they will know how to relate the rest of the narrative and the budget to what you want to do. If you cannot complete that first sentence, you are not yet ready to write a proposal.

A question that is rooted in theory is compelling because a good answer will satisfy a wider curiosity. Questions that are specific to particular times and places are merely interesting. Demonstrating that known principles operate in yet another place is helpful, but in the competition for scarce research funds, compelling questions win.

Virtually without exception, you will present a stronger case if your question is rooted in general ideas or in theory. You may wish to answer a question about a market in a Third World nation. If the answer does no more than enlarge our knowledge of the area and the nation in question, your question lacks generality, and it is unlikely that any agency will underwrite your research. If, on the other hand, your question about that market is informed by current ideas regarding the role of local markets in the development process, and if the answer bids fair to extend or change those ideas, your results will be generalizable to other places and hence of considerably greater interest to funding agencies.

Make sure that you can answer your question. Any scholar can pose a dozen questions of vital importance that cannot be answered given the constraints of current knowledge, methods, time, or money. Unless you can convince those who review your proposal that you can answer the question you pose, you will not be funded regardless of how important your question is.

The Research Plan

A research plan normally contains a scheme for gathering evidence or data that bear upon the question posed, a description of how the evidence will be evaluated, and closure between both the evaluated evidence and the question and between the project's results and theory. Make sure you can identify these components in your proposal.

Data. Leave no ambiguities regarding the kinds of evidence you will gather and how you will obtain it. If you propose survey work, include your interview schedule or questionnaire in an appendix. Ideally, you should already have tested the schedule or questionnaire to be confident it will yield the data you need. (If you cannot produce a schedule or questionnaire, you are not ready to submit a proposal.) If you intend to use data from government or commercial sources, identify those data precisely. General references to census, Department of Commerce, or Dun & Bradstreet statistics are insufficient.

Be certain your data can answer the question you have posed. Many reviewers will know as much or more than you do about survey data or standard data series. A surprising number of NSF applicants pose questions that cannot be answered by the intended data. Such blunders do not inspire confidence.

Methods. Tell reviewers precisely how you will evaluate and summarize the observations or data you gather. Whether your evidence is qualitative or quantitative, you are somehow going to meld your pieces of evidence into an answer to your question. Lay out that process clearly so reviewers can follow the sequential steps of your reasoning from raw data to the conclusions you will derive from those data. If you will use standard analytical techniques, specify them. Don't stop at general categories such as "content analysis" or "statistical techniques." Demonstrate that you are familiar with any questions or controversies regarding the methods you have selected. Neither your data nor your methods need be perfect, but you must convince reviewers that you are perfectly aware of any pitfalls inherent in using them.

Closure. First, build into your research plans closure between your question and your findings. Specify your criteria for answering the question you posed, and explain how you will determine whether your evidence-based conclusions satisfy those criteria. In short, make clear to proposal readers how you will know whether your project has succeeded or failed. (Be open-minded regarding failure. Proposals that can yield only one conclusion rarely fool reviewers. Negative results may be less satisfying than confirmed hypotheses, but they must be possible outcomes if you are engaged in research, as opposed to confirming tautologies.)

Second, build into your plan closure between your research results and theory. Remind reviewers of the general ideas that gave rise to your question, and explain how you will use the possible answers to confirm, alter, or enlarge the ways scholars think about the phenomenon you have studied.

This may be a good place to mention your plans for sharing your results. In most instances, that means publication; if your results have general or theoretical value, it means publication in refereed, scholarly journals. Dissemination via such outlets is a touchstone of basic research. If your underlying goal is not publication in a learned journal or a scholarly monograph, you should question whether you are proposing basic research.

The Budget and Narrative

Don't ask how much money you can have. You should have told the program officer how much you need in your preproposal discussions; if your need was out of line with program practice or resources, the officer will have so informed you.

How much *do* you need? Let the project determine the budget. Within the constraints of agency rules and guidelines, budget all direct and necessary costs of conducting the research you propose. If you cannot carry out the project without incurring the expenditure, include it. If an item would be nice to have but its absence would not compromise the integrity of the project, leave it out. Unless you have been given directions to the contrary by the program officer, do not select a target figure and then adjust your project accordingly.

One person's "direct and necessary" costs may seem like padding to another. Try this exercise: Imagine you are independently wealthy, so much so that you have become a patron of research. A colleague has asked you to underwrite a project and has just handed you the very budget you have just prepared. If you as patron would spend your own money on an item, the expenditure is probably justified. If you doubt that you would support a colleague's request for an item, eliminate it. If in doubt, ask an experienced colleague to evaluate your budget.

Explain every expenditure in your budget in an accompanying narrative that tells reviewers what the money will purchase and why it is necessary. Ambiguity is a vice anywhere in a research proposal, but especially in the budget. Expenses that are not documented are often questioned or eliminated. If you ask for salary, describe what you will do while you are being supported. If you need equipment, explain why, how it will be used, and where your cost estimates came from. Explain why and when travel costs must be incurred. Describe what research assistants will do while they are supported by project funds. Answer any question an intelligent, responsible reviewer or research administrator might ask about your budget.

Form and Format

I enjoined you earlier to follow each agency's instructions regarding proposal submission, and now I reiterate that advice. You may have a dozen good reasons for deviating from prescribed formats, but the overriding reason for hewing to them in every detail is that you may reduce your chances of winning an award if you don't.

Your proposal will loom large in your life, but it will be only one of several hundred or more that a program officer will handle in a year. Program officers, reviewers, advisory panel members, and agency administrators expect to find certain information in certain places and in certain order in the hundreds or thousands of proposals they process each year. If your honor demands creativity and independence in submitting your proposal, go right ahead and satisfy your honor. The warm feeling you derive may help you get over the disappointment you will feel if your proposal is returned or declined. Do it *their* way, not your way.

HALLMARKS OF SUCCESSFUL RESEARCH PROPOSALS

Winning proposals are self-contained, technically perfect, and exciting. Other things being equal, proposals that fail lack one or more of these qualities.

Anticipate Questions

Your proposal should answer any question a reasonable, qualified reviewer might ask. A proposal that anticipates and addresses valid reservations bespeaks a careful, committed scholar—the kind reviewers and program officers want to support. You need not resolve controversy over a topic, data, methods, or philosophical perspectives, but you must demonstrate that you are aware of relevant issues and how they might affect your results. Thus you must demonstrate familiarity with relevant literature and theory; you must locate your project in the context of that literature and theory.

Show that you have thought about all possible outcomes of your data collection, your analysis, and your results—in addition to the one that is dearest to your heart. When you have completed the first draft of your proposal, read it skeptically, and find as many conceptual, substantive, and methodological flaws as you can. Repair those shortcomings in all subsequent drafts. Ask a colleague to give your proposal a skeptical reading, and then incorporate the resulting suggestions.

Seek Technical Perfection

You will not convince reviewers and program officers that you are capable of handling big dollars if you do not attend to little details. Spelling, grammar, and arithmetic count—in fact, they count a great deal. You will damage your cause by submitting a proposal that is not technically perfect. Trivia, you snort? You are a deep thinker capable of great thoughts who cannot be troubled with minutiae? A proposal rife with spelling and grammatical flaws, erroneous formulae, and tables that do not sum causes reviewers to wonder whether its author will treat data and inferences with similar carelessness. You cannot afford to raise such doubts among reviewers. Make sure your proposal is neat, clean, and technically perfect. It is your only warrior in a brutal competition.

Good Writing Pays Off

Write your proposal the way you would write a love letter. Don't be afraid to express your enchantment and your passion for what you want to do. (If you feel neither, you have no business submitting a proposal. And don't substitute hype for the excitement you should feel; it won't work.) Reviewers respond favorably to genuine commitment. If you can convey the sense of wonder, excitement, and joy you should feel about your work, you will go a long way toward making reviewers want to underwrite your research. Write in the first person when appropriate, and write as clearly and as elegantly as you possibly can.

Remember, reviewers are busy people who are evaluating your proposal as a professional courtesy: make their task easy and enjoyable. Immediately after reading your proposal, a reviewer or panel member will write an evaluation. The way you have made your case will affect that evaluation. A proposal that engages interest because it is clearly and vividly written will put a reviewer in a kindly frame of mind. Tortured syntax and arcane vocabulary that make reading an agonizing chore will leave your reviewers testy.

Time invested in good writing will pay great dividends in agencies such as NSF that use advisory panels. An NSF panel member reads thirty to thirty-five proposals for each biannual meeting. If your proposal makes the part of the job for which you are responsible easier—or even a joy—you have gone a long way toward winning an award. You should allocate a few days solely to polishing your writing once you are satisfied with the substance and the technical details of your proposal.

PROPOSAL EVALUATION

Review procedures vary in detail among funding agencies. In general, most rely on some form of peer assessment of your proposal's merits in relation to the criteria by which the organization judges proposals. Peer recommendations and panels are normally advisory to a program officer who decides which proposals will be funded.

Peer Review

The NSF seeks as reviewers active researchers doing work as closely related to a proposed project as possible. The NSF geography program customarily requests comments from eight reviewers. Other programs and agencies use fewer; rarely does any agency use more than eight. Most agencies do not seek comments from individuals who stand to gain or lose if a project is funded or from close friends or known enemies of an applicant. A program officer may welcome nominations for reviewers or your opinions regarding reviewers who should be disqualified. Ask during preproposal discussions.

The NSF asks reviewers to comment on your qualifications to conduct the proposed research based on your previous accomplishments, on the intrinsic merit of the proposed project, on whether your results will be worthwhile, and on whether your project will augment the infrastructure of science. More simply put: can you do it, is it worth doing, will anyone use the results, and will science in general (as opposed to geography in particular, for example) be better off if this project is supported? The instructions other agencies give their reviewers are variations on these themes. Like many agencies, NSF requests written comments and a summary rating of each proposal ranging from excellent, through very good,

good, and fair, to poor. Ratings from excellent to good are implicit recommendations that the proposal be funded; a fair rating implies that funding is questionable; a poor rating means the reviewer recommends the proposal be declined.

Advisory Panels

Like most funding organizations, NSF uses advisory panels to evaluate proposals. Panel members write individual reviews and discuss each proposal at biannual meetings. Some proposals are dispatched quickly; others are discussed for an hour or more, as panel members identify and debate strengths and weaknesses in both the proposal and the peer reviews that it has generated. Discussion concludes when the panel reaches consensus, and consensus normally prevails; few of the six hundred or more proposals considered by panels during my tenure at NSF caused deep divisions in advisory panels.

Panels recommend that proposals be funded, declined, or funded subject to the availability of money. Proposals to be funded are clearly superior projects. Those recommended for decline are flawed or incomplete and will not be funded under any circumstances. The "fundable" category consists of meritorious proposals that are not as strong as those placed in the top category. If money remains after the top ones have been funded, some of those in the fundable category will be supported also.

Program Officer Decisions

Program officers execute the recommendations of reviewers and panels. They also administer the proposal intake and review process, select peer reviewers and panel members, and establish long-term funding priorities based on the advice they receive from their research communities and panel members. Most program officers (including those at NSF) have some latitude in executing recommendations from peer reviewers and panels. If they believe a proposal has been reviewed unfairly, they can fund proposals recommended for decline or decline proposals recommended for funding. Often a program officer will receive conflicting advice from different peer reviewers or from peer reviewers and the program's advisory panel. At such junctures program officers use their best judgment, weighing all the evidence available.

OUTCOMES

Proposals are either funded or declined based on the program officer's final recommendation. In either event, you will normally receive some explanation of the decision. Applicants to NSF receive verbatim copies of reviewer comments and a summary of the panel meeting discussion.

When Will You Hear? What Will You Hear?

During preproposal discussions, ask when decisions will be available. Find out whether you will be notified of the outcome as a matter of course or whether you should contact the program officer for a decision—and, if the latter, when you should call. Most officers welcome inquiries regarding proposals once decisions have been made.

Don't ever, unless specifically invited to do so, call a program director immediately after an advisory panel meeting. Running a two- or three-day advisory panel is exhausting, and program officers are semicatatonic the next day or two. In addition, it normally takes two weeks to achieve equilibrium between panel recommendations and available funds. Program directors have reserved special hot seats in Hell for applicants who call at 4:00 P.M. seeking information on proposals considered at a panel meeting that ended at 2:00 P.M.

Don't be angry if a program director tells you no more than the final decision on your application. You will be anxious for details, but the program officer has sixty or one hundred other applicants to inform. Your proposal will be in process somewhere, and no experienced program officer will talk about a proposal without documentation at hand. (It is too easy to confuse information from different proposals by relying on memory if you are handling a hundred proposals simultaneously.) In my experience, program officers are eager to serve their communities in any way they can. But they also experience pressures that are incomprehensible to academics. If you need reviewers' comments quickly, ask for them, but try to understand if they are not forthcoming instantly.

Good News

Congratulations are in order if you receive external funding. The competition is keen in any agency, and success is cause for satisfaction and celebration. There will be many details to attend to—some of them unpleasant, such as budget reductions, revised research plans, and perhaps an addendum to clear up ambiguities—but you should take pride in the vote of confidence you have received from your peers.

Before charging off to the field, the laboratory, or the library, however, read the proposal reviews thoroughly and with an open mind. If your program officer selected reviewers wisely, the reviews will contain constructive suggestions on ways your project can be improved and warnings about possible pitfalls. Good advice from eight or so experts in your specialty should not be ignored. Revise your research strategy and plans in light of this expert advice.

Start thinking ahead to your obligations as a grant recipient. The NSF expects scholarly publication to be the major outcome of its awards. Refine your research plans with that end in mind. You may receive one research award on the basis of promises; you are unlikely to get another unless results of your previous award(s) have been published in refereed journals. Now is the time to think about your next research proposal: How will you demonstrate that your research produces worthwhile, publishable results? How can your current project serve as a foundation for subsequent work?

Oh Woe!

Declinations hurt. You labored honestly and earnestly on your proposal, yet your peers deem your best effort unworthy of support. What shame! Moreover, it is public shame. Your department chairperson, your research office, and perhaps some of your colleagues know that you've been turned down. Some of the reviews may contain infuriating comments. Put it all aside for a few days or weeks until the pain, shame, and anger, which are normal reactions to declination, subside.

When the suffering has lessened, take a cold-blooded, hard-headed look at the reviews and your proposal. What summary ratings do you have? If most are very good or good, you may want to revise and resubmit your proposal. If most are good and fair, you probably have an inherently weak proposal and should begin anew with another question.

Read the written comments carefully to determine which identify real flaws in your proposal and which result from reviewers' misunderstandings of what you intended to do. What is the tenor of the reviews? If most dwell on serious flaws in a neutral or negative manner, resubmission is unwarranted; try another idea. Are most of the criticisms constructive? Does the panel summary hint at or invite revision? If so, submit a revised proposal after discussing it with the program officer. Revisions that honestly address flaws identified in earlier submissions succeed more often than first-time submissions.

What you receive, even if it is not the money you asked for, is valuable advice. Make good use of it. If reviewers identified flaws in your thinking or your plans, repair them to the best of your ability. If reviewers misunderstood your proposal, rewrite it so they will understand, even if you think some of them are fools. It is your responsibility to explain everything so clearly that even fools will not mistake your intentions.

BE PERSISTENT

Most proposals submitted by most applicants fail. The Geography and Regional Science Program at NSF funded only one out of five of the regular research proposals received during my tenure. The number of scholars in geography and related disciplines who succeed more than once in every three attempts can be counted on one hand. What distinguishes those who win external funds from those who fail is that winners are willing to fail. If you need external funding for your research, you must accept failure as a price of success. Unless you are exceptionally talented, you must be willing to fail more often than you succeed.

Winning external funding is a learnable skill. It is less developed among geographers than in disciplines with longer and stronger traditions of external funding. A handful of graduate programs teach the skill formally; most geography graduate programs, however, do a poor job of preparing their students to compete for research funds. If you were or are being trained in one of the latter, you will have to teach yourself. Whether you are taught formally or on your own, you must remember that you can't win unless you try; you surely won't receive if you never ask.

SUMMARY

The starting point for your search for external funds must be a clean, tractable problem that you are driven to solve. Without that anchor, you will drift helplessly. Tackle one problem at a time, and make it a tightly circumscribed problem. During my tenure at NSF I handled three or four proposals that were deemed too narrow to be funded, but hundreds that were not funded because they were hopelessly ambitious. The applicant could not have completed them in a career, let alone in a year or two.

Preproposal contact with the funding agency and with the person who will handle your proposal is not of questionable propriety; rather, it is essential. The modal proposal at NSF arrives on the program director's desk with no preproposal discussion. Such proposals often contain fatal errors, which could have been avoided had the applicant done a little homework and then talked to the program director for ten minutes. Preproposal discussions are proper and desirable. Your time and effort are precious. Nobody wants to see them squandered on hopeless causes. Make sure you have selected an appropriate agency and an appropriate problem before you write your proposal.

Do everything in your power to make your proposal complete, self-contained, and elegant. Weed out contradictions and ambiguities ruthlessly, and obtain help from colleagues in doing so. Read your proposal aloud to yourself. If you find yourself wondering what you are saying, rewrite the proposal until it is absolutely clear and precise. It will be sent to first-rate, responsible scholars. It may also fall into the hands of a few knaves and fools. Program officers try to avoid such reviewers, and they discount their reviews, but they are not omniscient. Write a proposal that even knaves and fools will applaud.

Be persistent. Don't expect to fail, but be willing to fail. If you are not, you will never surmount what is most likely to be your first and last experience in seeking external funding. The modal NSF applicant submits a proposal without preproposal contact, is declined, and never submits another proposal. Thus the modal applicant does not profit from the experience and the advice received in response to the first application; the entire exercise is a waste of everyone's time. You fell the first few times you tried to ride a bicycle, and if you ever struggled with a musical instrument, you probably remember how hard you had to work to play a simple tune. It is unreasonable to expect instant success in obtaining external funds.

Winning external funding is a learnable skill. If you are reasonably bright, energetic, sane, have a reasonably good idea, and are persistent and willing to learn from your failures, you can win funding for your research. Somewhere in the world, someone wants to fund the research you want to do. You will find that agency if you search long enough. You will convince that agency to underwrite your work if you are committed to research and willing to learn from your mistakes.

ON ACADEMIC SURVIVAL

On the Way Toward Tenure and Promotion

Leonard Guelke

Tenure and promotion are critical elements in a successful academic career in the United States and Canada. To be fully independent, a university professor needs the job security tenure provides. Tenure is fundamental to academic freedom and scholarly independence, providing individuals with institutional protection to hold unpopular views—in short, to be intellectually independent. Tenured faculty members have an obligation to take advantage of this privileged position. The independent and free university is as important to a democratic society as a free press or independent judiciary, and tenure is a crucial element in guaranteeing this independence and freedom.

Compared to tenure, promotion is of less importance. The security and the nature of the job do not change with promotion, and in many institutions salary is also unaffected. Thus, promotion, unless it is tied to tenure, is largely a symbolic recognition of achievement. Yet, most people consider recognition important even if it makes no real difference to their work. Promotion from assistant to associate professor has a somewhat different meaning than promotion from associate to full professor. In the former case promotion demands less and is often associated with tenure. It symbolizes the transition from probationary status to that of a permanent and established member of the university community. This promotion is an important career objective for all faculty members. The promotion from associate to full professor is less crucial, since some professors will remain in the associate ranks for most if not all of their careers. It is not an assured step in everyone's career, but the vast majority of faculty will eventually be promoted to full professors, and such promotion is considered to be an important achievement.

The steps necessary, for a beginning faculty member who has been appointed to a tenure-track position to become a tenured, full professor, are, for the majority straightforward and painless. Occasionally, however, the process is anything but smooth, and individuals can get caught up in situations where they become victims of academic in-fighting or political tensions within a department or university. Although these situations are the exception rather than the rule, individuals who are beginning their academic careers can take precautions to protect their positions.

This chapter provides some general advice for those seeking tenure and promotion. The conditions and rules operating at the many centers of higher learning in Canada and the United States vary widely, and new faculty will have to acquaint themselves with their local environments. The point to emphasize is that faculty members can help their careers by keeping themselves fully informed on local procedures and the expectations of the institution in which they work.

GETTING STARTED

In today's world of tight academic budgets and uncertainty about future funding, new faculty members are frequently appointed on definite-term contracts. Many universities favor definite-term contracts because they provide financial flexibility, and in a financial crisis the contracts of definite-term employees need not be renewed. An individual has no security of tenure under this kind of contract even if the job is described as a tenure-track position. A first objective of a new faculty member on a definite-term contract should be a probationary appointment, which does provide some measure of security. Chairpersons will often be sympathetic to the situation of junior faculty on definite-term appointments, and that sympathy should if possible be converted into administrative action.

An important task of newly appointed faculty members is to gain some idea of what is expected of them. If a person appointed to a definite-term position receives an indication from his or her chairperson that some refereed publications and evidence of teaching effectiveness would help in securing a probationary appointment, priorities should be set accordingly. Wherever possible, an individual should obtain expectations and implied promises in writing. Written comments are worth something; verbal assurances, in practice, mean very little even when the individual making them is perfectly sincere.

New faculty members should protect their interests by diplomatically asking for any implied agreements to be set down in writing. If circumstances make this difficult, they could send their chairperson a letter that includes something along the following lines: "Further to yesterday's conversation, I would like to confirm that you expect it will be possible to convert my definite-term contract to a probationary appointment on the publication of two refereed papers and evidence of my teaching effectiveness." This sentence might be followed by some details of a plan of action to meet these goals. By getting such understandings in writing, new faculty members will protect their positions by making the criteria on which they will be assessed as clear as possible. They will also make it difficult for the administration to create more stringent criteria in the future. For example, a new chairperson would be placed in a difficult position repudiating the expectations of achievement of his or her predecessor, even if the former chairperson's expectations were considered too low.

There will be many demands placed upon newly appointed faculty members' time, and they must be careful in setting priorities. The main emphasis should be on teaching and publication. If a doctorate has not been completed, much of the new faculty member's energy should be devoted to its completion. Today it is virtually impossible to have a successful career as an academic without a doctorate, and the longer its completion is postponed, the more likely it is that it will never be completed. Once the Ph.D. has been completed, a new faculty member should aim at publishing his or her research, either as a book or as a series of refereed articles. In looking for appropriate publication outlets a faculty member should consult with colleagues and, if possible, get their advice and critical comments on draft manuscripts.

The publication demands—both of volume and of quality—placed upon new faculty members will vary widely depending on the nature of the institution. Whatever the demands might be, a new faculty member should endeavor to reach them in a planned and systematic way, by identifying academic goals and outlining steps for achieving them. Even if individuals feel that the standards of their institution are very high, they should work out ways to attain them. A defeatist attitude toward one's own career prospects is likely to become a self-fulfilling prediction. The challenge of achieving a level of scholarly excellence depends on dedication and hard work. The successful academic is as much a product of hard work as the successful athlete. True, there is a measure of natural talent in both enterprises; but the differences between levels of achievement in academia are largely a function of varying degrees of individual commitment and self-discipline in working toward desired goals.

Although publication is crucial to an academic career, teaching is of equal importance in most universities. The beginning faculty member should try and devote large amounts of time to both activities. As a teacher and faculty member, the former graduate student will need to reorder personal priorities. In graduate school the emphasis is on advancing knowledge, often on narrow and well-defined fronts. Teaching undergraduates, however, demands an entirely different focus. The subject matter must be placed in a broad context and made intelligible to students with general backgrounds. New faculty members should discard the idea that their students will have as much interest in a topic as faculty do. The task of the professor is to make the subject matter as interesting as possible, using techniques that have more in common with the theater than academia.

New instructors should not wait until the end of a term to find out how successful they have been. They should give out class tests, ask students questions, and keep alert for any signs of dissatisfaction. An instructor should be prepared to change tack if it becomes clear that the teaching style or course content is not coming across well. Sometimes professors have difficulty simplifying material and thus add too many qualifications to every statement they make. New teachers should remember that they are teaching undergraduates, often at an elementary level, not presenting a lecture to their thesis examining committee with the expectation that any broad statement will be challenged and evidence of its validity demanded.

Most universities have provisions for an annual performance review, and these reviews will provide promotion and tenure committees with evidence about an untenured faculty member's contribution to the university. In discussing their progress with their chairperson, untenured faculty members should discuss performance in terms of progress toward tenure and promotion. If the chairperson expresses satisfaction with an individual's performance and agrees that the faculty member is making good progress toward tenure and promotion, it is wise to request that this statement be added to the evaluation. If problems are identified, a faculty member should insist that they be as narrowly based and explicitly defined as possible. For example, vague statements about teaching problems should be challenged and the exact nature of any real or alleged problems defined. New professors may find, for example, that their rigorous and demanding courses have generated a great deal of hostility from students. This could indeed be a problem, but the way it is described in an evaluation is important. The positive achievements should be emphasized; and the problem should be located within a context that emphasizes the willingness of the professor to make appropriate changes.

In subsequent evaluations, problems that have been addressed successfully should receive appropriate positive acknowledgement. It may be necessary for a faculty member to point out explicitly achievements that have gone unnoticed, and to have them recorded on the annual evaluation. If new faculty members are at an institution with no formal annual review mechanism, they should seek to ensure that there is some regular record of their contribution to the university.

Above and beyond the research and teaching requirements of a university career is the requirement to be a good colleague. New faculty members should take their time to get the measure of the institution in which they are working and should seek to "fit in." Although new faculty members may find some aspects of their working environment disagreeable, they should be cautious about expressing strong views. Instead, they should be agreeable and accommodating—but not too agreeable and certainly not unduly deferential. If you look as if you have no views of your own and do everything you are told to do, even when the demands are unreasonable, you could be perceived as weak. On the other hand, if you show contempt for colleagues and the ways of the institution, you are likely to be resented, and there will be strong pressure from many colleagues to deny tenure regardless of your academic achievement. New faculty members should get involved in their departments in a positive and constructive way, but their main emphasis should be on establishing themselves as scholars and teachers.

THE TIMING OF TENURE AND PROMOTION

Faculty members who, from first appointment, have given attention to the objectives of gaining promotion and tenure will be well placed to judge when they might be ready to make formal application. If the chairperson has indicated satisfaction with the progress and contribution of a faculty member, the question can be raised of whether formal application should be made. If the chairperson is encouraging and feels the time is right to apply, the candidate can usually go ahead with some confidence. However, if the chairperson advises waiting a year, it will probably be wise to do so. The chances of making a successful application for promotion and tenure with lukewarm or no support from your chairperson are very poor. If the chairperson seems uncomfortable discussing prospects, the candidate should press him or her for some definite answers. There can be few experiences more devastating than to discover, after a negative decision, that one had taken for granted the support of a "silent" chairperson whose silence indicated no support at all. If the chairperson suggests at annual evaluation time that you will be ready to apply for tenure and promotion next year, ask that this be put into your annual evaluation. Although having to wait an additional year may be frustrating, there is little point in putting forward applications with slim chances of success. An early application for promotion and tenure is usually judged on severer criteria than those coming forward at the normal or required time.

THE APPLICATION

Once faculty members have decided, or are required by policy, to make application for tenure or promotion, they should ensure that all evidence of career accomplishments is assembled in a convenient form. The curriculum vitae, or c.v., is critical. This document is a summary of an individual's education and experiences, and it should reflect them accurately and in a way that makes the candidate look as good as possible. The c.v. should be organized so as to conform to any local regulations and should contain accurate and clear details of degrees, courses taught, students supervised, committees served on, and publications, all arranged in a way that brings out their scholarly significance. The c.v. must be well-organized and well-presented, without errors of punctuation and spelling. A sloppy document will suggest a sloppy candidate.

In addition to a c.v., the application should include a brief statement of career objectives and highlights of academic achievements. This document is more than a summary of a c.v. Rather, it seeks to show there is an underlying coherence to a career and points in the direction of future research efforts to be built upon earlier accomplishments. The career summary can place a candidate's work in perspective and help the promotion and tenure committee make sense of a more detailed c.v. The objective of candidates should be to present their work as a focused research endeavor within a clearly defined disciplinary context.

An important part of most tenure and promotion proceedings is the solicitation of letters from external referees. The establishment of the referee list is of major importance, because the opinions of the referees carry great weight in tenure and promotion deliberations. The use of external referees ensures some outside control of the tenure and promotion process and is designed to guarantee the integrity of the procedure. Candidates should seek to ensure that the referees selected to review their work have the necessary knowledge and philosophical orientation to do it justice. Humanist geographers, for example, would not want their work evaluated by a dedicated positivist or by a radical geographer. A referee should have a broad appreciation of the subdiscipline within which the candidate has worked, and a good knowledge of the regional setting in which research has been undertaken.

Candidates should be knowledgeable about the people working in their field, be able to help suggest appropriate referees (if this freedom is allowed), and be able to challenge individuals who for one reason or another might make questionable referees. It is far easier to insist on eliminating a name before any letter of reference has been written, than to seek to question negative appraisals of your work from a referee already acknowledged as an expert in a field. In looking at lists of referees, a candidate should make use of friends and colleagues, sounding them out about the credentials and fairness of the people who might be approached. If it is not possible to have a referee removed from a list, candidates should express their reservations about the individual in a letter to the chairperson.

NEGATIVE DECISIONS

The first reaction of candidates who have been rejected by a committee of their peers is usually anger and disbelief. These emotions are entirely understandable and are often fully justified. Anger and disbelief, however, are poor guides to effective action. Unsuccessful candidates for promotion or tenure must keep their heads and look for ways to accept or appeal the decision. Appeals based on questioning the integrity and motives of the promotion and tenure committee will seldom achieve positive results, but will have the undesirable side effect of alienating colleagues.

A candidate in receipt of a negative decision should seek advice and guidance immediately from a trusted colleague or faculty association adviser. An adviser who has knowledge of the university and how it works is absolutely essential if appropriate and effective action is to be taken. This is not to suggest that candidates should not get fully involved in their own appeal, but someone not emotionally entangled in the situation is needed to offer advice and to weed out emotional hyperbole in correspondence.

The first decision to make is whether or not to appeal. This decision will be tied very much to the context, and to the reasons for the negative decision. In cases where candidates are not at risk of losing their jobs, they might be well-advised not to appeal on a first application for tenure or promotion. In making this decision the candidate should study carefully the letter or report of rejection. If such a letter is not volunteered by the committee chairperson, the candidate should insist upon it—if necessary, with the help of the local faculty association.

The letter of rejection should contain reasons why tenure or promotion has been denied and explicitly suggest ways in which identified shortcomings might be corrected. The candidate should make sure the demands are specific and attainable within a reasonable time period. If the candidate is satisfied on this account, it probably makes sense not to appeal. In future applications the candidate should be in a position to show that professional accomplishments have met the criteria noted by the promotion and tenure committee that rejected the earlier application. In this way there is some insurance against having to face future situations where a new committee, unaware of earlier applications, demands higher levels of achievement.

If there is doubt about what is required of a candidate, if a candidate considers the committee has made unreasonable demands, if a candidate is rejected for a second or third time, or if a candidate has run out of time and must be tenured or dismissed, then appeals are usually justified and, in some cases, absolutely essential as a matter of career survival. Yet, no matter how strongly a candidate may feel about a situation, it is critical that all actions related to an appeal be carefully considered, strategic decisions.

APPEAL STRATEGY

The objective of an appeal is to find grounds on which to challenge an original decision. These grounds can take several forms and must be advanced in the context of the appropriate university policies and guidelines. In searching for grounds for appeal, an individual should look for ways in which a reversal of an original decision can be justified, either by the group involved in making the original decision or by a different body. Once a decision has been made, unless it can be shown that the original decision involved some error, there is generally reluctance to change it. Most academics have strong respect for the intellectual judgment of colleagues, and in a one-to-one contest between candidates' judgment and that of a committee, candidates will lose in almost every case—even if they are clearly intellectually superior to the members of the committee. Universities are built on the respect for scholarly judgments. And, in the opinion of most academics, this principle is essential, even sacrosanct, for the proper functioning of the university itself.

As the first appeal gets under way the applicant should make sure that a file of all relevant material is collected, which includes copies of all letters written or received. This file will establish the sequence of events and provide a written record of what has happened. These letters will become the basis of any future appeals and possible legal action.

A first ground for an appeal is procedural. The policy of each university usually outlines in some detail the procedures to be followed in tenure and promotion cases. The candidate in receipt of a negative decision should scrutinize the policy and make sure it has been followed. It is surprising how many times official procedures are ignored and how sloppy some chairpersons and deans can be in forming committees. A decision appealed on the basis of a serious procedural error (or errors) should lead to the original decision being canceled.

Any decision made under an incorrect evaluation of the criteria of judgment would fall within the category of procedural errors. Thus an overzealous chairperson might insist that a candidate for tenure or promotion possess an outstanding teaching record when, in fact, university policy specifies that a capable scholar need only be a competent or good teacher. Although the bases on which a judgment has been made are not always available to a candidate, the letter informing a candidate of a negative decision usually makes them pretty clear. In appealing, the candidate can draw an appeal committee's attention to this kind of discrepancy and provide it with the grounds necessary to reverse the earlier decision.

The concept of equity is the second ground for appeal. If candidates can show that they have been judged in a way that is different from other colleagues in similar positions, this can be grounds for appeal, but it is often difficult to prove such differential treatment. The easiest cases are those that can be clearly documented. For example, if candidates are rejected for promotion or tenure on the grounds that they have not taught large classes well, but documentation exists to show they have, in fact, better ratings in such classes than a recently promoted or tenured colleague, the negative assertion can be countered effectively.

An appeal can be very effective if a promotion and tenure committee has presented hearsay evidence in support of a decision. Such evidence is inadmissible.

Thus, in a letter outlining a negative decision, the chairperson might refer to complaints he or she heard from students about a particular course. These kinds of statements do not constitute evidence and should never be used in a promotion and tenure judgment. The candidate certainly can appeal any decision where it is evident that hearsay or questionable evidence has been used.

The third strong ground for appeal is the production of new evidence. In many respects this is one of the very best options, making it possible for a committee to reverse a decision without raising questions about the fairness of the original decision. New evidence can be presented in a variety of ways. The candidate might receive information that a refereed manuscript has been accepted for publication in a major journal. This evidence may be exactly what is needed to firm up a scholarly record that may have been assessed as a little thin before. In another instance, candidates might discover that material pertinent to their case was not circulated or seen by committee members and would wish to draw it to their attention.

One or all of the above grounds might be used in a particular appeal. In writing appeal letters, make sure that they are based on a written judgment. Never write an appeal letter on the basis of a verbal communication. If a letter containing the promotion and tenure committee's negative decision does not provide the basis on which a decision was made, the first thing candidates should do is write a letter requesting the reasons for the negative judgment. It is impossible to object to a judgment if the basis on which it was made is kept a secret. Once candidates have received a written judgment with reasons for denial, they should prepare a response in consultation with an adviser. The response should be kept reasonably short and to the point. There will be an urge to respond to every single line. Avoid this temptation, and focus on the main issues in terse, unemotional language. The response letter will be part of a paper trail that might be needed at many levels of appeal. It should be used to enhance a case, by showing the candidate to be a reasonable, logical person—not as a desperate individual lashing out in all directions.

The principle of freedom of information should be used wherever possible to obtain reference letters that might have been used by a committee to make a decision. Any factual errors in such material should be pointed out immediately, as should any special circumstances. For example, the impact of a negative letter from a student about a candidate's teaching might be reduced or nullified by pointing out that the student missed many lectures or labs and ended up with a D in the course.

Applicants will likely need considerable tenacity to take an appeal through all the internal committees and hearing mechanisms available at a particular university. However, the basic aim of the appeal is to review the candidate's case. It may be difficult to achieve such a review at the lower levels of the administration, from colleagues who work together on a day-to-day basis and who have influenced each other's views in subtle or not-so-subtle ways. In the context of universities that are organized on department, faculty, and university levels, the candidate might not achieve a fresh and fully independent review until the appeal reaches the university level. In fact, the existence of an appeal procedure at different levels is explicit recognition that mistakes can occur.

The candidate who exhausts all the appeals allowed by tenure and promotion policies can still, if unsatisfied, take further action. In some institutions it is possible to launch a grievance against officers of the university administration on the grounds that the appeal process was handled unprofessionally or inappropriately. The justification for such a grievance would focus heavily on the procedures employed and their possible impact on the fairness of the decisions reached. If a grievance is not possible, legal advice may be sought privately or through the auspices of a faculty association. The courts can neither force a university to give a person tenure nor require it to promote someone. However, courts are interested in fairness and due process. In certain situations they can intervene to ensure that an individual's right to be treated fairly and equitably is upheld.

CONCLUSION

The futures of new faculty members are largely in the hands of others. But, by ensuring that they know the rules and are able to put the best light on their accomplishments, faculty members can have a major influence on those who make decisions about them. They should plan their careers in ways that seek to ensure that their accomplishments will meet the requirements of tenure and promotion at their institution.

How Does Service Relate to Tenure and Promotion?

Robert A. Muller

The question of service relative to promotion and tenure primarily concerns assistant professors who will need to earn promotion and tenure. All my academic experience has been in departments with graduate programs, but in terms of service, I don't think there is much difference between Ph.D.-granting departments and departments that offer M.A. or M.S. programs as a terminal degree. There are, however, great differences among universities, among the various colleges and academic departments within the same university, and even within a given department as the administration of that department and the college continue to change. Nothing in the academic world seems to be stable and unchanging forever.

Most faculty are judged for promotion and tenure according to a three-part standard: teaching, research, and service. Over the last ten years the hierarchy has been research first, teaching second, and service last; a cynic might well claim that the weighting is 85 percent research, 10 percent teaching, and 5 percent service. The same cynic also might claim that research is measured strictly by counting articles in refereed journals and by one's success at winning grants and contracts.

Unfortunately, the cynic is sometimes not far from the truth, and tenure at some of the research universities depends mostly on an acceptable level of publication in refereed journals. Teaching effectiveness may be judged solely on the basis of a computerized student questionnaire and/or student gossip passed on by other faculty. And service may well not be considered at all. My personal estimate of the overall relative weighting of research, teaching, and service is roughly 50, 30, and 20 percent, respectively. This weighting suggests that it is nearly impossible for a young assistant professor to be promoted and receive tenure without some significant record, and further promise, of publication in refereed journals. It is possible to be promoted with a mediocre teaching record, but promotion is highly unlikely when the research record is weak. Once in a while poor teaching will be overlooked because of a truly outstanding record of publication or grantsmanship. Service is rarely the critical issue because usually an outstanding service record will not offset a weak record of publication.

With the weighting leaning so heavily toward research and publication, departments can become populated with faculty who have relatively little concern for student and classroom needs, to say nothing of the service functions of the department and university. Given the primary demands on the faculty, they focus first on their own research priorities. Consequently, the departments do not function very smoothly as organizational units. The necessary chores—the teaching and service needs—often are conducted inefficiently and there is little long-term stability.

An obvious improvement, in my view, is to recognize that most of the faculty are neither equally interested nor equally able in terms of their research, teaching, and service commitments. The larger departments could then develop a mix, with several faculty members dedicated to excellent teaching and working closely with students. Perhaps they would find the need to publish a paper, say, once in five or more years. Publication would hardly matter, of course, because they would earn promotion and tenure in terms of their classroom and student performance. The department also would need one to three members especially committed to administration and service, and they could be judged primarily on these responsibilities.

This idealized department, therefore, would include a carefully cultivated faculty mix. Scholarly professors would spend most of their time with graduate students in research and publication. Teaching faculty would be concerned mainly with the undergraduate classes and especially the students, with increased attention to the successful development of the young people and their educational needs. And finally, there also would be a few who administer the departments and the programs, with special attention to the service needs of the department and university.

SERVICE IN THE DEPARTMENT

Some graduate departments try to encourage new assistant professors to do research and to publish by limiting teaching loads and service commitments in their first semester or even their first year. But there are many chores and duties in a graduate department that cannot be performed by the department chairperson alone; sooner or later new faculty have to begin taking turns with some of the responsibilities.

My personal view is that assistant professors should avoid at all costs department committees that make judgments on the performance, merit, or salary of colleagues. As an assistant professor on such a committee, you will get into trouble and make enemies, regardless of your fairness and wisdom, and you will be especially vulnerable if you do not have tenure. Also, the time necessary for serving on these committees can be enormous and open-ended. Nevertheless, at some institutions you may be encouraged or even pressured to sit on one of these committees because university or college guidelines require representation from all levels of the faculty.

Department curriculum committees usually offer fewer hazards and possibly even some rewards for innovative programs. I assume that new assistant professors arrive armed with the latest techniques developed in their graduate programs, and with fresh memories of the lively debates over the pros and cons of various research methodologies. Therefore, they can help to improve the research and teaching programs of their new departments by adding "state of the art" modifications to existing programs and perhaps introducing new programs. There is, of course, the danger that a new assistant professor can be too zealous with new approaches and ideas, especially when proposals appear to threaten the position, work, or graduate students of an established faculty member. You need to proceed with care and caution; an intemperate young assistant professor rarely moves smoothly toward promotion and tenure.

Another department committee that has a threatening side is the one that makes decisions about graduate applications and, especially, assistantships and fellowships. No matter what the guidelines say about eligibility and semesters of support, there are always special cases deserving of consideration, and other cases where the personal interests and wishes of senior faculty predominate. Again, this type of committee service offers little of practical value for the untenured young faculty member, along with the real danger of confrontation over whose graduate students should receive the limited financial support available. Probably, untenured faculty members ought to rely on the good will of more senior faculty to assure that they receive their fair share of the graduate student support.

Perhaps I have dwelled too much on the negative possibilities when serving on some department committees. However, most departments have other committees that offer opportunities for service duties, but without most of the hazards. One that comes to mind is registration duty. At some universities this means spending time during registration at a central facility where students walk through a portion of the registration process. This chore normally is divided up so each faculty member is responsible for only a half day, or perhaps a full day. So, your duty is only one day a semester, and you get to meet students, learn a bit about the structure of their programs, and sometimes meet interesting faculty from other departments.

One of the most time-consuming and challenging chores of all can be serving as the undergraduate adviser to the department's majors. Obviously, this is a very important responsibility, especially in terms of recruiting a reasonable number of able students into the program each year; some departments may separate the recruitment and advising functions. The adviser also needs to have the experience and wisdom to develop the best program of electives for each student, and to know when and how to grant the occasional exceptions that are necessary from time to time.

This advising function should be delegated to faculty members who can balance sensitivity to the academic needs of all the students against the "social needs" of a few students. Some students will want to be with the adviser as much of the time as possible for all sorts of reasons. Some of them will wish to be around interesting people of an earlier generation; others may want to talk about issues; and still others simply may want to get your attention. The point is that the demands for time will be open-ended, and students will take all of the time that you allow. Incidentally, the same should be said for relationships between individual instructors and their students.

When this service function is delegated to assistant professors fresh out of graduate school, there is often the danger that, with student experiences still etched in their memories, they will relate more to the students on a social level and less to faculty colleagues. Over the years, all too often I have heard about promising young faculty who committed so much time to students that there simply was not enough time left for research and publication. Because the advising function requires experience, judgment, and so much time, in almost every case it should be left to tenured faculty.

Although it is hardly considered a service function, I wish to add a few remarks about advising graduate students and serving as their major professor. I have always believed that the development of well-trained, successful graduate students is one of the most important functions of Ph.D.-granting departments. From this point of view, being the major professor for a bright graduate student, or being a member of graduate committees, is a very important service function to the discipline, the department, and the individual faculty member. On the committee, or as the major professor, you have opportunities to interact with colleagues and the graduate student and in some situations to learn about the "highs and lows" of leadership responsibilities. I see this service function in terms of learning how to grow in the academic community, as one that has the potential to provide a healthy record of success.

Other department chores are often organized on a service or committee basis. Examples are duties or committees for department equipment, room allocations and other space needs, library purchases, and so on. Most of these chores are not terribly demanding, and they often provide opportunities to learn in a minimum of time the many dimensions of an academic department.

A less common type of duty is "service" as the department chairperson. Normally, this position would not be offered to an assistant professor without tenure, but I have seen this happen several times in emergency situations. I do not consider here extended-period chairperson positions, which should be thought of in terms of university administration rather than service. These positions are sometimes labors for a lifetime; at other times they represent upward steps on the administrative ladder of a university.

In departments where the tenured associate and full professors are expected to work through a rotation as the department chairperson, the position should be viewed as one of open-ended service. It is a very difficult position, and the chairperson is usually caught somewhere between the needs of the administration, the individual faculty of the department, and the students. All too often there can be basic conflicts among these needs.

In general, I think that too much is expected of a chairperson in these times of complex administrative procedures. During a typical three-year tenure, the chairperson essentially has to put research and publication on hold, or at least far down on the priority list. The chairperson faces so many conflicting demands that it is difficult to convince individual faculty members that the person is doing the best job. Hence, the chairperson rarely has that good feeling of a job well done. It is, therefore, the most challenging and demanding service of all at the department level.

SERVICE FOR THE COLLEGE

Service on college committees is more demanding, in terms of understanding and appreciating a much broader range of detail and the problems that encompass as many as ten to twenty academic disciplines. Committee meetings can be tedious, since individual members often have their own academic agendas, which are sometimes in conflict with your department's or even your with own principles. Again, it would probably be prudent for new assistant professors to put off college service, especially the more demanding committees, until you have your research and publications under way. But you should use this grace period to begin to learn about the formal and informal networks among the departments of the college and, for that matter, with other colleges as well. You should keep alert to the potential for service on a very specialized college committee where you can use and further develop your own particular training and skills.

For the untenured assistant professor, the ideal college service could well be on some special-purpose committee organized either by the dean's office or, sometimes, as a committee of a college senate. A general example that comes to mind is a committee charged with improvement of student writing skills; another might be one organized to develop student computing skills throughout the college; a third and more delicate example is a committee whose task is to upgrade geographic knowledge among the students of the college. You will find that there are many opportunities, and it is sometimes possible to match your special interests and skills with the charge of a particular committee. In this service situation, you ought to enjoy the duties, learn something of the structure and networking of the college, and begin to develop a record of helpful and useful service. At the opposite extreme is assignment to a committee with which you have no common interests. To avoid this, you would be well-advised to find out what is available, and even to volunteer for service on a committee of your choice.

Most colleges have two or more very important committees: (1) courses and curriculum and (2) promotion and tenure. The latter is usually restricted to tenured faculty at the associate professor level or higher, and there is no need to discuss it here. But the courses and curriculum committee—or whatever it might be called—often includes members from the ranks of the untenured faculty. My experiences are that this committee is trying and demanding, and it is normally better served by tenured faculty with more than just a few years of experience.

Membership in a college senate is another form of service, and attempts are often made to encourage election of senators equally from the three ranks of faculty—assistant, associate, and full professors. Perhaps there is an element of prestige for untenured assistant professors to be elected to the senate. Some college senates are very active, and participation can mean a lot of excitement. Some college senates work closely with college administrators, but there are other instances where the senators may have an adversary relationship with college administrators. Needless to say, the latter situation is not one for untenured assistant professors. My advice, again, is to work into these kinds of service duties after promotion and tenure.

Those who have a Ph.D. tend to share a collective disdain and distrust of administrators at all levels. The status of a department chairperson varies from university to university. At some institutions, a chairperson is treated as part of the administration, at others, as a member of the faculty doing temporary administrative duties. And there is a difference between faculty members who take a three-year turn as chairperson, and those who serve for ten to twenty years or more.

Some faculty discover early on that they enjoy administrative work more than teaching, research, and publishing. Some will therefore drift into administration by taking on more and more service duties, until eventually they are appointed to administrative posts. Others make a deliberate decision to work into administration, often by seeking initial appointments as assistant or associate deans.

My point here is not that administration is a bad route to follow, but that you should enter it with a firm foundation of academic credentials. I don't mean you need to be well-founded and successful in research and publication before becoming an academic administrator; it seems to me that such positions require very different personal skills. Rather, my point is that if you should have notions about turning toward administration, you can begin with service committees at the department and college levels. But you will be a much more effective administrator later on if you have gained the academic credentials of a tenured associate professor, and preferably of a full professor, before you take that final step into administration as an assistant or associate dean. In most cases, a "part-time" appointment as assistant or associate dean means spending forty or more hours each week working on administrative matters, with little time left over for further research and publication.

SERVICE FOR THE UNIVERSITY

Virtually everything I have written about service to the college can be applied equally to university service. The committee activities may be no more time-intensive than similar college committees, but the scope of interests and the necessity for some general understanding will often be even more comprehensive than at the college level. I would suggest avoiding these committee activities until research themes, publications, and promotion and tenure are all firmly under control.

Each university is organized somewhat differently, but most governing structures include a faculty senate with the members elected by the faculty, and graduate and athletic councils with membership either by appointment or by election. These official bodies offer interesting and even exciting opportunities for service. But, for the untenured assistant professor, the distractions from research and publication could be self-destructive. And, because of the competitive nature of college athletics today, membership on the athletic councils may represent the most hazardous service duty that can be performed in American universities! I firmly believe that one should have tenure before undertaking service duty on these senates and councils.

There are, however, many university-level, special-purpose committees that may provide service opportunities for untenured assistant professors. Again, it is a matter of matching your skills and interests with an appropriate committee—one that provides a much-needed service in an orderly fashion without overwhelming demands of time. Most universities have so many of these committees that I won't attempt to list examples here.

Unfortunately, there is one special committee that should be avoided by untenured assistant professors: the faculty grievance committee. Avoiding this committee can be difficult, particularly because university bylaws often require that it include more than token representation from all faculty ranks, from both sexes, and from minority groups. Hence, there can be strong administrative pressure on untenured assistant professors to sit on this committee.

I believe that, although faculty grievance committees are necessary and vital, they are hardly the place for assistant professors who have a limited number of years to earn promotion and tenure. The sessions can be long, tedious, and very trying, and it is sometimes difficult not to become emotionally involved with marginal situations that are, indeed, complex. I know of productive faculty members who claim to have lost an entire semester of research over a single such case. This is hardly the appropriate service responsibility for a person charged with earning promotion and tenure mostly through research and publication.

SERVICE TO THE COMMUNITY

Community service can range from an occasional lecture outside the university to a professional service component with a separate university budget, which charges fees for selected services provided to users beyond the university campus. The common denominator is that the service is provided to or for individuals or groups outside the direct university community.

In most situations community services are barely recognized in terms of promotion and tenure. The general exceptions would be wherever there is an allocated budget for extension work. Today, many agricultural and home economic units have large budgets for extension services, and individual faculty in those units normally have some proportion of their time allocated to extension activities (in some cases this may be as high as 100 percent). In the extension situation, what we normally think of as service becomes an officially designated work responsibility and is most likely evaluated for promotion and tenure. But I know of very few budgets for extension services within geography departments.

Most university administrations nevertheless recognize the need for a service component in the total university effort, and my own experiences with hundreds of faculty members over the years suggests that many, but not all, geographers enjoy talking formally about aspects of their work to the interested public beyond the university. Obviously, the nature of one's research is a factor here. Some faculty conduct research that is so technical and specialized that it would be difficult to bring it to the public; I suppose there are instances where hardly anyone would care if they did. But there are few research topics in geography that are not of interest to a proportion of the public outside the university community.

Service situations may offer opportunities to develop special programs that can evolve into long-term career efforts, combining nearly all of the components of the university mission of teaching, research, and service. It is possible to begin as an untenured assistant professor, but it would be more prudent to attempt such an endeavor after tenure has been earned.

I will conclude this section on service to the community with two examples from my own department at Louisiana State University (LSU). LSU has a combined department of geography and anthropology, with graduate programs in both disciplines. I will present, in brief, one example in applied anthropology and another in applied geography.

A number of years ago, when I was chairman of the department, we hired an outstanding young physical anthropologist in a "soft money" position, with funding guaranteed for only an initial year, because he recognized the unusual opportunities for research and service in this state. From his graduate training he brought special skills in forensic anthropology; specifically, he identified selected vital statistics from human skeletal remains. The Louisiana State Police and other local law enforcement agencies needed this kind of support, and the services were provided on a volunteer basis. This service effort required an exhausting commitment, from one to three days at a time, on a basis of more or less instantaneous demand.

After several successful identifications, the service effort received significant media attention, and the university administration allocated tenure-track funding for the position for the following year. A new tenure-track position was thus created out of this "service effort," leading to the successful development of a program and training for graduate and undergraduate students. The faculty member in question earned promotion and tenure a few years later with an excellent record of publication.

The second example is personal: it concerns the climate information services program that I established at LSU about fifteen years ago. I was a full professor with tenure at the time, so this example stands in contrast to the anthropology situation described above.

In a budget-cutting measure in 1973, the National Weather Service closed down all of the federally funded state climatologists' offices in each of the states. Their duties had been to maintain the climate records of each state and to provide current information services to federal and state agencies and the general public. The state climatologist for Louisiana had been located within our department, and he turned over his files to me when he left. After his departure, a climate data user had to turn to the National Climatic Data Center (NCDC) in Asheville, North Carolina, for help and information. But NCDC was neither staffed to handle the deluge of requests nor well-acquainted with the geography of Louisiana. As a result, it was not possible to provide the same interpretations relative to the environmental and economic impacts of climate.

In 1973 I began to provide a bit of "free" service over the phone from my office. After a few years the requests for data and interpretations increased to the point where I had to go to state agencies and the university for staff support to continue. To simplify a long and complex account, the information services program evolved into a separately funded Office of State Climatology—a unit of our department. The activities generated new courses in climatology, captured the interest of undergraduate students, attracted funding for graduate students, and gave rise to new, tenure-track climate positions in our department and in the department of agricultural engineering. Again, a service opportunity evolved into a department program with interrelated elements of research, teaching, and service. Moreover, I see these as two examples where service allowed faculty to develop their own special niches within a department structure.

One more service needs to be discussed here: service to the discipline and to professional societies, in our case especially the Association of American Geographers (AAG) and the Canadian Association of Geographers (CAG). On the national scene, the AAG and CAG stand alongside other great academic associations, but both the AAG and the CAG can be only as strong as the collective commitment of the geographers across the two countries. Therefore, service to geography as a discipline and specifically to the AAG and CAG is a necessity.

Service to the AAG and the CAG means participation in the various functions and activities of the associations. Especially important are participation in the programs of the annual meetings and the specialty groups, committee and journal responsibilities, and at least some publication in the journals of the associations. There is little question in my mind that the "prosperity" of individual geography departments in colleges and universities depends in some measure on the extent to which the AAG and the CAG can project a vital and useful image of the discipline. Hence, this element of service is critical and needs to be developed at the very beginning of the academic career, if it has not already been initiated during graduate studies.

CONCLUSIONS

Service in the university, whether at the department, college, or university level, is a vital component of the university structure. Universities depend on this gratuitous faculty service in order to function. Without these services administrative costs would be significantly higher at all levels.

I believe the system of services within the university is essentially good. It forces faculty to step out of their research "cocoons" to become aware of the broader needs and concerns of units within the university and of the university as a whole. In addition, service—especially to the community beyond the university—can lead to rewarding opportunities with potential for department growth and special-interest programs.

If I have seemed a bit negative in this essay, it is only because of the pressure imposed on junior faculty for research and publication in refereed journals in a short time period. Assistant professors have many conflicting challenges to face in only a few years. They have to begin teaching new courses and seminars, hopefully with adequate time allocated for interacting with the students. They must extend research and publication beyond their dissertations—for the first time out from under the wings of their major professor. Many need to write proposals for grants and contracts. They have to cope with a new university and a new town. And most must also cope with family members who surely will have their own set of adjustments to make in new communities.

Given all of these circumstances, and given the relative weighting of service for promotion and tenure, I have to conclude that untenured assistant professors should avoid service duties as much as possible, especially the big consumers of time and emotions. But they always should remain alert to the possibilities for serving with a minimum commitment of time and where there is potential for professional and departmental support.

The Geographer as Administrator: Perspectives on Survival for Geography Departments

Risa Palm

Many geographers have chosen to become administrators as well as faculty members. Geographers-as-administrators can have a major impact on the health of geography departments in their own institutions as well as at other colleges and universities. Administrators have many opportunities to invest or disinvest in departments: changing the number of faculty members available to units, changing the resource base of the administrative units, setting priorities for new facilities among competing departmental interests, and changing entrance or exit requirements that affect relative departmental student enrollments. In addition, through conversations with administrators in comparable positions from other institutions, geographer-administrators can promote the cause of geography departments elsewhere.

Program review is the principal means by which major resource re-allocations are made within the university. In this chapter, I describe the program review process at the University of Colorado, a process that is now common in major universities throughout the country, and recommend ways in which geography departments can prepare themselves to attain favorable outcomes of a review process.

Program review is the periodic review of academic departments, wherein decisions are made about expansion, the establishment of new programs, or even the termination of departments. As the former chairperson of the Program Review Panel at the University of Colorado, and as a member of several external review teams at other institutions, I have had the opportunity to observe programs prospering from this process, and also to see programs suspended or terminated as a result of program review. In this essay, I will discuss strategies for surviving and even benefiting from the inevitable program review.

THE PROGRAM REVIEW PROCESS

The University of Colorado instituted a process of regular program reviews in 1980. Most major universities now have such a regular process, and departments or programs expect a major periodic review every seven to ten years. Several books and journal articles have been written on the subject of how to conduct a program review (Harpel 1986; Barak 1982; Seeley 1981; Wilson 1982), and administrators are trained to carry out these reviews in conferences conducted by organizations such as the National Center for Higher Educational Management Systems.

A fairly standard format for program reviews has evolved. These involve the following four steps:

1. The production of a self-study by the unit reviewed.
2. The evaluation of the unit by a group of faculty and students from other related departments within the institution. In the case of geography, such faculty and students may be drawn from geology, economics, sociology, political science, anthropology, atmospheric sciences, and so on.
3. An evaluation of the unit by an external team of one to three persons in the same field (that is, geographers from other colleges or universities).
4. An administrative integration of the internal and external reviews.

A description of each of these steps follows, along with recommendations for maximizing their effectiveness.

The Self-Study

Every unit being reviewed is asked to prepare a self-study describing the members of the department, its contributions to the college and the university, and an assessment of its problems and needs. The university's office of institutional research (or its equivalent) provides the department with data on the number of students enrolled in classes, numbers of majors and graduates, support budgets, and other information.

Before beginning the writing of the self-study, the department should check on the accuracy of the information provided by the university office of institutional research or planning. Although these data are believed to be accurate by the institution, it is possible—and even common—for them to contain errors. Among the sources of error are computerized counts of majors that may omit double-majors (listing only the first of the two or more majors); assessments of the numbers and dollars of external grants and contracts to be credited to the department, particularly when members of the department are also members of organized research units (such as institutes) that may receive sole credit for external grants; and simple errors resulting from the miscoding of data. Since there is often little time to check these figures during the process of preparing the self-study, it is imperative that the unit maintain accurate counts on such data at all times. This should be the responsibility of the department chairperson, although it may be delegated to a staff assistant.

The self-study should be both descriptive and analytical. Regardless of the guidelines provided by the college or university for preparing the self-study, the geography department self-study should provide a description of the following four items: (1) the nature of geography and the role of this particular geography department within American geography; (2) the history of this particular department; (3) the contributions made by the department to the mission of the college or university; and (4) the strengths and weaknesses, as well as opportunities for improvement, of the department in carrying out its mission. Let me elaborate on each of these topics.

First, the geography department self-study should discuss the nature of geography and the contribution that this particular department makes in the context of American (or Canadian) geography, largely because geography is a relatively little-known field to administrators. Administrators tend to be drawn from the faculty, and many have been department chairpersons of major units such as chemistry, English, psychology, or physics. These administrators frequently received their doctorates at the most prestigious universities in the country, many of which do not even have a geography department. As a result of their own training, therefore, many administrators are unfamiliar with collegiate geography and the research conducted in our field. They do not know of either the major research advances within geography in recent years or the relationships between geography and other fields of knowledge represented within the college. They may not know that remote sensing or geographic information systems are actually integral topics within a geography curriculum. They may not know that sophisticated forms of location theory and locational analysis are the realm not of economics but rather of economic geography. They may not understand the centrality of spatial relationships or environmental interactions to the field of geography. They may not understand that the subjects of natural or technological hazards or environmental change fall within the realm of geography. Therefore, it is important to spell these ideas out and to present a brief synopsis of current research in the field.

The prior familiarity that administrators have with geography is to a great degree a function of their previous experience. If a new dean or vice chancellor arrives on campus directly from a university with a strong and visible geography department, it is probable that the administrator will have some awareness of the strengths and potentials of the field. On the other hand, administrators who are themselves products of undergraduate institutions where geography was not present—including institutions such as Brown, Princeton, Yale, Harvard, or Stanford, all of which lack an undergraduate geography program—and who have had administrative experience in colleges or universities with weak or nonexistent geography programs are very likely to have little or no familiarity with the field.

Second, it is important that administrators understand the history of geography departments, primarily because this history may help to explain the concentration of departments on limited numbers of topics. If the department was formerly merged with geology, there may remain either a strong emphasis on geomorphology in the department or a complete absence of physical geography. If the department was formerly joined with anthropology, there may be a particular emphasis on cultural geography in the present department. The evolution of the department, culminating in its independence, has implications suggesting contributions of this discipline to the educational and research mission of the institution. These independent contributions should be highlighted within the recapitulation of the history of the department, where appropriate.

Third, listing the contributions the department makes to the mission of the college or university is the most important element in the self-study. This should be a topic for consideration in any planning sessions or retreats held by the department. In a teaching institution, where large numbers of student credit-hour production and high quality teaching are the primary institutional mission, it is important that the geography department contribute to this mission in a significant and visible way. In a research institution, where research prestige is most valued, the geography department must also emphasize excellent and recognized research. In institutions that value the fact that all of their departments rank in the top twenty or top ten of their respective fields, the geography department should rank among the top departments in research reputation. In institutions that stress the numbers of dollars in grants received by faculty members, the geography department, too, must perform well on this criterion. In short, a geography department providing large numbers of student credit hours, but with relatively little relative research strength, is in a good position in a teaching institution but in an extremely vulnerable position in a research institution. Similarly, a humanistically oriented geography department is clearly vulnerable in a college where garnering external research dollars is a value stressed by the administration.

How can one determine the institutional mission? I will grant that examining the rhetoric of the university administration does not give a clear answer. While it is obvious that the primary mission of the community college and some state colleges and universities is teaching, the stress that some state colleges and universities place on research for salary increments, promotion, and tenure belies the unique importance of a teaching mission. Furthermore, it is likely that administrators in even the most "hard rock" research institutions—those that place extreme emphasis on external grant awards garnered, national honors awarded to faculty, and research reputation—may state that their institutions give equal weight to research and teaching. Let no one be fooled by such protestations. At the University of Colorado, for example, admissions to several graduate programs were suspended despite moderate to heavy demand for courses in these departments and healthy numbers of majors per faculty member. Admission was suspended not on the basis of teaching excellence—indeed, several of the faculty members in one of these departments had won university-wide teaching awards—or student credit-hour productivity at the graduate or undergraduate level. Instead, in every case, the decision was based on the conclusion that the faculty were not producing research of sufficient quality to maintain the program.

If there is any doubt about the orientation of your college or university, it is useful to look at its public relations materials—the programs or aspects of programs it sells to prospective students and faculty members. Does such material claim an emphasis on individualized or innovative instructional methods? Does it feature particular research programs? Does it recount the rankings of departments with respect to external grant funding or reputation of the faculty in scholarly work? Another way to determine the institutional mission is to investigate the experience of other departments that have undergone reviews. On what bases have they been enhanced in size or suspended? An analysis of the values of the college and its administration is well worth an investment of time in order to orient the departmental mission to those activities most valued by the institution.

Fourth, the self-study should contain an analysis of both the strengths and the weaknesses of the department, as well as an assessment of its needs. Of course, the department will wish to show the ways in which it contributes to the education of students (majors and nonmajors), provides service to the institution, and advances the status of geography itself. Faculty contributions should be mentioned in terms of teaching (numbers of students taught, numbers of majors, placement of students in advanced degree programs or in jobs, and comments from alumni about the quality of the education received from the department), research (numbers of publications, assessment of impact of publications, honors won by faculty members, and reputation of the faculty with respect to research and scholarly work), and service (university and professional service, including participation in faculty governance). To sustain a case regarding research, teaching, and service, it is terribly important for departments to keep accurate data, even when not required to do so by the institution. The department should keep track of alumni, both for the purpose of indicating to the administration success in placement of graduates and also to keep alumni informed of the activities of the department so that their support may be enlisted when needed. The department should maintain tabulations of honors garnered by faculty and students, grants and contracts awarded to department members, and other indicators of success.

The weaknesses of the department should also be stated frankly, in the context of the kinds of remediation that would shore up these weaknesses. The reason for candor in this portion of the self-study is the fact that weaknesses cannot be hidden in the review process, and any attempt to cover them will result in a distrust of the unit and even greater scrutiny by the internal faculty review team and the administrative team.

It is very important that the self-study be the product of the entire department, and that all faculty members participate in its preparation and agree with its assessment. Although consensus on this document may take longer to achieve than the production of a self-study by the chairperson and the executive committee, it is important that faculty members feel that they have fully participated in the assessment and evaluation of the department.

On the other hand, once the self-study is completed it is very important that all of the faculty close ranks on any possible sources of disagreement in the document. The review teams and the administration will look for signs of internal rifts, and faculty should resist the temptation to use the review process as an opportunity for self-serving pleas to outside reviewers or the administration. It is important to remember that the administration looks for signs of mutual respect among faculty members, even among those who do not agree philosophically or professionally. Such mutual respect is sometimes as important in the positive evaluation of a department as any other index of performance. Some evidence to support this assertion is found in the histories of geography departments that have been closed by administrators—not because of the lack of quality of the faculty, but rather because of internecine battles that, in light of their consequences, clearly should have been suppressed in the interest of the stability of the department and the welfare of the discipline.

Internal Review

Universities and colleges differ in the processes they use for enlisting review teams. At the University of Colorado, there are three stages of review: an Internal Review Team, an External Review Team, and a final review by a standing committee.

The Internal Review Team is composed of five members from within the university but outside the department. It usually includes three faculty members, one graduate student, and one undergraduate student, all from cognate departments. The reason for such a team is to involve a faculty review process by people familiar with the university context but outside the department under review. The Internal Review Team reads the self-study and interviews faculty, students, and alumni to assess the strengths, weaknesses, and needs of the department. It usually provides the most critical of the set of reviews, since the faculty participants are well aware of the constraints and opportunities within the institution and are committed to high standards of quality in their own departments and others.

The Internal Review Team submits a report outlining its findings about the department, along with a set of recommendations for administrative action. The department is given an opportunity to comment in writing on any errors of fact (not opinion) contained in the report of the Internal Review Team, and this departmental response is included along with the report at all subsequent stages of the review. The report, along with the self-study, is sent to the members of the External Review Team, and is also considered by the standing committee.

External Review

Consultation with external reviewers is a common aspect of program reviews in many universities and colleges. The reason for inviting external reviewers is so that university administrators, who may be unfamiliar with a particular discipline, can garner an assessment of the nature of this discipline and the status of the university department within this field.

At the University of Colorado, as elsewhere, the External Review Team is composed of distinguished faculty members from the same discipline but from other universities. The team is selected by the standing committee based on a combination of (a) nominations submitted by the department and (b) recommendations from professional associations. Members of the External Review Team are selected on the basis of their national prominence in the field; they are usually past presidents of professional associations, members of the National Academy of Sciences, editors of prominent journals, or other individuals who have achieved particular distinction in the discipline.

Since the department plays some role in the selection of the External Review Team through its nominations, it is important that these nominations be made carefully. The department should avoid nominating individuals who are graduates of the department or former faculty members, or who in any other way could be seen to have a conflict of interest. Any hint of departmental manipulation of the review team, or even an indication that the department is willing to be reviewed by individuals who are less than nationally prominent, will compromise the confidence of the internal faculty and the administration in the standards of the department and its full participation in the review process.

For geography departments, the External Review Team is likely to provide a sympathetic review, with recommendations to the administration that more resources be directed to the department. It is in the department's best interests, therefore, that the sources of these recommendations be individuals whose opinions will influence the administrators responsible for the geography department and the program review. From the perspective of the administration, the most valuable part of the visit of the External Review Team is the opportunity to meet with distinguished faculty members from the discipline from outside the institution, to gain an unbiased perspective on the research frontiers of the discipline, and to receive an assessment of the role the local department plays in the national picture of the profession.

The report of the External Review Team usually contains both findings and recommendations, although its format is less standardized than other aspects of the review process. The report is submitted to the department for comments on issues of fact (not evaluation or opinion), and it is forwarded to the Program Review Panel for the last part of the process.

The Program Review Panel

At many universities, the External Review Team report is simply forwarded to the university administration for action. At the University of Colorado, however, there is yet another step in the review process: an analysis of all of the reports by a standing committee called the Program Review Panel. The panel is a group of eight faculty members selected by the vice chancellor for academic affairs in consultation with the faculty governance organization (the Boulder Faculty Assembly), as well as one graduate student and one undergraduate student. It is chaired by the associate vice chancellor for academic affairs (who does not have a vote). Each faculty member on the panel serves as a liaison to one department being reviewed that year and is responsible for meeting all reviewers and producing a first draft of the final report and recommendations.

The report of the Program Review Panel consists of two parts: findings and recommendations. The report is based on information supplied in the self-study, the report of the Internal Review Team, and the report of the External Review Team, as well as testimony from appropriate deans (e.g., arts and sciences, and the graduate school). The final report is debated by the panel and is the end product of the program review. This is the only aspect of the program review process that results in a public document; the self-study and the reports of the Internal and External Review Teams are confidential, shared only with the department, its dean(s), the Program Review Panel, and other university administrators who have a "need to know" the contents of any of the documents. The report of the Program Review Panel is forwarded to the vice chancellor for academic affairs and, when accepted, becomes available to the regents and the general public.

The Program Review Panel report is usually presented as a set of findings and recommendations. Findings are statements of fact as well as assessments of quality. They concern the following:

1. The faculty and its teaching, research, and service accomplishments
2. The students and the impacts of the departmental experience on them
3. The curriculum offered in the department
4. The service offered by the department to the college, the university, and the profession
5. The facilities of the unit

Recommendations range widely: from recommendations concerning the construction of new facilities or remodeling, to the addition of new faculty in particular areas; and from issues of departmental governance and administration, to curriculum revision. These recommendations by the standing faculty committee are taken very seriously by the administration, since they are the result of careful study and reflection by a series of informed faculty reviewers. New departments have been formed, while other departments have been closed or had their admissions suspended. Thus, this process is crucial to the health and future well-being of a department.

SUMMARY

Periodic program review has been adopted widely by colleges and universities throughout North America as a means for (1) systematically reviewing and evaluating existing programs and (2) making administrative decisions about investment or disinvestment. I have indicated several ways in which a department may either prosper or falter in the course of such reviews. My recommendations concerning the review process can be summarized and reiterated as follows:

1. Be sure to explain the nature and context of geography as a discipline in the self-study in order to educate administrators unfamiliar with geography as to recent technical and theoretical advances and research frontiers.
2. Pay close attention to the overall mission of the university, and orient the department to this mission. Then, when the program is reviewed, stress the part the department plays in achieving the institutional mission.
3. State both the strengths and weaknesses of the department in the self-study, as well as ways in which modest investments in the unit could achieve major improvements.
4. Involve all faculty in the preparation of the self-study.
5. Try to ensure that, once the self-study has been completed, all of the faculty join ranks to support the conclusions contained therein.
6. Recognize that departments that are governed well, with faculty who respect one another and whose mission is consonant with that of the university or college, have nothing to fear in a program review. Indeed, such a review can be viewed as an opportunity to present a detailed case of accomplishments and needs to the administration, which at that particular moment has its attention focused on the geography department.

The program review can be understood as a threat or an opportunity. A defensive posture by the department in response to a review perceived as a threat is unlikely to yield positive results. Attempts to cover up problems are likely to backfire, as shrewd colleagues probe the inadequate defenses. On the other hand, program review can be an opportunity to call attention to the needs and aspirations of the department. If the review is approached positively and is conducted with rigor and candor, the review process can be used to promote the goals and interests of departments.

Barak, R. J. 1982. *Program Review in Higher Education*. Boulder, CO: National Center for Higher Education Management Systems.

Harpel, R. L. 1986. The anatomy of an academic program review. Association for Institution Research (AIR) Professional File Paper No. 25.

Seeley, J. A. 1981. Program review and evaluation. In *New Directions for Institutional Research,* No. 31, ed. N. L. Poulton, 45–60. San Francisco: Jossey-Bass.

Wilson, R. F. 1982. Designing academic program reviews. In R. F. Wilson (ed.) *New Directions for Higher Education,* No. 37. San Francisco: Jossey-Bass.